Ramanujan's Forty Identities for the Rogers-Ramanujan Functions

Memoirs
of the
American Mathematical Society

Number 880

Ramanujan's Forty Identities for the
Rogers-Ramanujan Functions

Bruce C. Berndt
Geumlan Choi
Youn-Seo Choi
Heekyoung Hahn
Boon Pin Yeap
Ae Ja Yee
Hamza Yesilyurt
Jinhee Yi

July 2007 • Volume 188 • Number 880 (first of 4 numbers) • ISSN 0065-9266

American Mathematical Society
Providence, Rhode Island

2000 *Mathematics Subject Classification.* Primary 11P82, 11F27, 33D15.

Library of Congress Cataloging-in-Publication Data
Ramanujan's forty identities for the Rogers-Ramanujan functions / Bruce C. Berndt ... [et al.].
  p. cm. — (Memoirs of the American Mathematical Society, ISSN 0065-9266 ; no. 880)
  "July 2007, volume 188, number 880 (first of 4 numbers)."
  Includes bibliographical references.
  ISBN 978-0-8218-3973-7 (alk. paper)
  1. Number theory. 2. Combinatorial identities. 3. Functions, Theta. 4. Generating functions. 5. Partitions (Mathematics). I. Berndt, Bruce C., 1939– II. Title.
QA241.R26 2007

2007060759

Memoirs of the American Mathematical Society

This journal is devoted entirely to research in pure and applied mathematics.

Subscription information. The 2007 subscription begins with volume 185 and consists of six mailings, each containing one or more numbers. Subscription prices for 2007 are US$649 list, US$519 institutional member. A late charge of 10% of the subscription price will be imposed on orders received from nonmembers after January 1 of the subscription year. Subscribers outside the United States and India must pay a postage surcharge of US$38; subscribers in India must pay a postage surcharge of US$43. Expedited delivery to destinations in North America US$53; elsewhere US$130. Each number may be ordered separately; *please specify number* when ordering an individual number. For prices and titles of recently released numbers, see the New Publications sections of the *Notices of the American Mathematical Society.*

Back number information. For back issues see the *AMS Catalog of Publications.*

Subscriptions and orders should be addressed to the American Mathematical Society, P. O. Box 845904, Boston, MA 02284-5904, USA. *All orders must be accompanied by payment.* Other correspondence should be addressed to 201 Charles Street, Providence, RI 02904-2294, USA.

Copying and reprinting. Individual readers of this publication, and nonprofit libraries acting for them, are permitted to make fair use of the material, such as to copy a chapter for use in teaching or research. Permission is granted to quote brief passages from this publication in reviews, provided the customary acknowledgment of the source is given.

Republication, systematic copying, or multiple reproduction of any material in this publication is permitted only under license from the American Mathematical Society. Requests for such permission should be addressed to the Acquisitions Department, American Mathematical Society, 201 Charles Street, Providence, Rhode Island 02904-2294, USA. Requests can also be made by e-mail to reprint-permission@ams.org.

Memoirs of the American Mathematical Society is published bimonthly (each volume consisting usually of more than one number) by the American Mathematical Society at 201 Charles Street, Providence, RI 02904-2294, USA. Periodicals postage paid at Providence, RI. Postmaster: Send address changes to Memoirs, American Mathematical Society, 201 Charles Street, Providence, RI 02904-2294, USA.

© 2007 by the American Mathematical Society. All rights reserved.
Copyright of this publication reverts to the public domain 28 years after publication. Contact the AMS for copyright status.
This publication is indexed in *Science Citation Index*®, *SciSearch*®, *Research Alert*®, *CompuMath Citation Index*®, *Current Contents*®/*Physical, Chemical & Earth Sciences.*
Printed in the United States of America.

∞ The paper used in this book is acid-free and falls within the guidelines established to ensure permanence and durability.
Visit the AMS home page at http://www.ams.org/

10 9 8 7 6 5 4 3 2 1   12 11 10 09 08 07

Contents

Chapter 1.	Introduction	1
Chapter 2.	Definitions and Preliminary Results	5
Chapter 3.	The Forty Identities	7
Chapter 4.	The Principal Ideas Behind the Proofs	15
Chapter 5.	Proofs of 35 of the 40 Entries	25
Chapter 6.	Asymptotic "Proofs" of Entries 3.28 (second part), 3.29, 3.30, 3.31, and 3.35	85
Chapter 7.	New Identities for $G(q)$ and $H(q)$ and Final Remarks	93
Bibliography		95

Abstract

Sir Arthur Conan Doyle's famous fictional detective Sherlock Holmes and his sidekick Dr. Watson go camping and pitch their tent under the stars. During the night, Holmes wakes his companion and says, "Watson, look up at the stars and tell me what you deduce." Watson says, "I see millions of stars, and it is quite likely that a few of them are planets just like Earth. Therefore there may also be life on these planets." Holmes replies, "Watson, you idiot. Somebody stole our tent."

When seeking proofs of Ramanujan's identities for the Rogers–Ramanujan functions, Watson, i.e., G. N. Watson, was not an "idiot." He, L. J. Rogers, and D. M. Bressoud found proofs for several of the identities. A. J. F. Biagioli devised proofs for most (but not all) of the remaining identities. Although some of the proofs of Watson, Rogers, and Bressoud are likely in the spirit of those found by Ramanujan, those of Biagioli are not. In particular, Biagioli used the theory of modular forms. Haunted by the fact that little progress has been made into Ramanujan's insights on these identities in the past 85 years, the present authors sought "more natural" proofs. Thus, instead of a missing tent, we have had missing proofs, i.e., Ramanujan's missing proofs of his forty identities for the Rogers–Ramanujan functions. In this paper, for 35 of the 40 identities, we offer proofs that are in the spirit of Ramanujan. Some of the proofs presented here are due to Watson, Rogers, and Bressoud, but most are new. Moreover, for several identities, we present two or three proofs. For the five identities that we are unable to prove, we provide non-rigorous verifications based on an asymptotic analysis of the associated Rogers–Ramanujan functions. We think that this method, which is related to the 5-dissection of the generating function for cranks found in Ramanujan's lost notebook, is what Ramanujan might have used to discover several of the more difficult identities. Some of the new methods in this paper can be employed to establish new identities for the Rogers–Ramanujan functions.

Received by the editor on January 28, 2005.

1991 *Mathematics Subject Classification*. 11P82, 11F27, 33D15.

Key words and phrases. Rogers–Ramanujan functions, Ramanujan's lost notebook, theta functions, Rogers's method, modular equations, transformation formulas, asymptotic expansions of Rogers–Ramanujan functions.

Research partially supported by grant MDA904-00-1-0015 from the National Security Agency (Berndt).

Research supported by grant R01-2003-000-11596-0 from the Basic Research Program of the Korea Science and Engineering Foundation (Y.–S. Choi).

Research partially supported by a grant from the Number Theory Foundation (Yee).

CHAPTER 1

Introduction

The Rogers–Ramanujan functions in the title are defined for $|q|<1$ by

(1.1) $$G(q):=\sum_{n=0}^{\infty}\frac{q^{n^2}}{(q;q)_n} \quad \text{and} \quad H(q):=\sum_{n=0}^{\infty}\frac{q^{n(n+1)}}{(q;q)_n},$$

where here and in the sequel we use the customary notation $(a;q)_0:=1$,

$$(a;q)_n:=\prod_{k=0}^{n-1}(1-aq^k), \quad n\geq 1,$$

$$(a;q)_\infty:=\lim_{n\to\infty}(a;q)_n, \quad |q|<1.$$

These functions satisfy the famous Rogers–Ramanujan identities [31], [25], [27, pp. 214–215]

(1.2) $$G(q)=\frac{1}{(q;q^5)_\infty(q^4;q^5)_\infty} \quad \text{and} \quad H(q)=\frac{1}{(q^2;q^5)_\infty(q^3;q^5)_\infty}.$$

At the end of his brief communication [26], [27, p. 231] announcing his proofs of the Rogers–Ramanujan identities (1.2), Ramanujan remarks, "I have now found an algebraic relation between $G(q)$ and $H(q)$, viz.:

(1.3) $$H(q)\{G(q)\}^{11}-q^2 G(q)\{H(q)\}^{11}=1+11q\{G(q)H(q)\}^6.$$

Another noteworthy formula is

(1.4) $$H(q)G(q^{11})-q^2 G(q)H(q^{11})=1.$$

Each of these formulae is the simplest of a large class." Ramanujan did not indicate how he had proved these two identities, which, as we shall see below, are two from a list of forty identities involving $G(q)$ and $H(q)$ that Ramanujan had compiled.

In his paper [32] establishing ten of the identities, Rogers remarks, "these [identities] were communicated privately to me in February 1919 ..." Rogers did not indicate if further identities were included in Ramanujan's communication to him. We remark that Ramanujan returned to India on February 27, 1919, and that the short paper [26] was recorded in the minutes of the London Mathematical Society on March 13, 1919. Thus, both the paper to the London Mathematical Society and the letter to Rogers were evidently sent only days prior to Ramanujan's departure.

In 1933, Watson [34] proved eight of the identities, but with two of them from the group that Rogers had proved. Watson confides, "Among the formulae contained in the manuscripts left by Ramanujan is a set of about forty which involve functions of the types $G(q)$ and $H(q)$; the beauty of these formulae seems to me to be comparable with that of the Rogers–Ramanujan identities. So far as I know, nobody else has discovered any formulae which approach them even remotely; if

1

my belief is well-founded, the undivided credit for the discovery of these formulae is due to Ramanujan." This last statement appears to be so obvious, especially since the manuscript was evidently in Watson's possession, that one wonders why he wrote it.

Ramanujan's forty identities for $G(q)$ and $H(q)$ (which do not include (1.2)) were first brought in their entirety before the mathematical public by B. J. Birch [11], who in 1975 found Watson's handwritten copy of Ramanujan's list of forty identities in the Oxford University Library. Ramanujan's original manuscript was in Watson's possession for many years and now has evidently been lost. Watson's handwritten list was later published along with Ramanujan's lost notebook [29, pp. 236–237] in 1988. Certain pairs of the identities are linked, and so it is natural to place them, in fact, in 35 (not 40) separate entries.

D. Bressoud [14], in his Ph.D. thesis, proved fifteen from the list of forty. His published paper [15] contains proofs of some, but not all, of the general identities from [14] which he developed in order to prove Ramanujan's identities. All the proofs of Rogers, Watson, and Bressoud employ classical means, although it would seem that in most cases the proofs are not like those found by Ramanujan.

After the work of Rogers, Watson, and Bressoud, nine remained to be proved. A. J. F. Biagioli [10] used modular forms to prove eight of them. At this moment then, only one of the forty identities has not been proved by any means, but it is clear that modular forms can be used to establish this last identity. About such proofs, Birch [11] opines, "A dull proof would have little value – in fact, all the functions involved in the identities are essentially theta functions, so modular forms of known level with poles of bounded order at known places, so the identities may presumably be verified by just checking that the first hundred or so powers of x are correct." It should be remarked that Biagioli's [10] proofs are more elegant than one might discern from Birch's remarks, for Biagioli used Fricke involutions and other properties of modular forms to drastically reduce the number of terms envisioned by Birch. In fact, in most cases, Biagioli required only a few terms.

In this paper, we offer proofs of 35 of the 40 identities. Some of the proofs that we present were found by either Rogers, Watson, or Bressoud. However, most of the proofs presented in this paper are new. Frequently, we provide two or three proofs of an identity. Our goal has been to find proofs for all forty identities that Ramanujan might have given himself. Indeed, in several of our proofs, we utilize modular equations found by Ramanujan and recorded in his notebooks [28]. Although all the proofs offered here are in the spirit of Ramanujan's mathematics, it is to be admitted that for some proofs, knowing the identity beforehand was a distinct advantage to us in finding a proof. It is unfortunate that we have failed to find proofs of five of the identities, four of which were proved by Biagioli [10] using the theory of modular forms and one of which has not been proved at all. However, for each of these five identities, we offer heuristic arguments showing that both sides of the identity have the same asymptotic expansions as $q \to 1-$. It is very likely that Ramanujan discovered many of his identities for $G(q)$ and $H(q)$ by examining asymptotic expansions. Ramanujan was an expert on asymptotic expansions, and in his last letter to G. H. Hardy written on January 12, 1920, Ramanujan discussed the asymptotic expansions of his new mock theta functions and compared them to the asymptotic expansion of $G(q)$ with which he opened his letter [8, p. 220]. In the last section of this paper, Section 7, we state several new identities involving the Rogers–Ramanujan functions; proofs may be found in [**9**].

In conclusion, we think that modular equations were central in many of Ramanujan's proofs, and we also think that Ramanujan may have discovered some of his identities by examining the asymptotics of theta functions and the Rogers–Ramanujan functions. Although some of our proofs may be those found by Ramanujan, it is clear that all of us, including the aforementioned authors and the present authors, have not unveiled some of Ramanujan's principal ideas, which remain hidden by an impenetrable fog.

CHAPTER 2

Definitions and Preliminary Results

We first recall Ramanujan's definitions for a general theta function and some of its important special cases. Set

$$(2.1) \qquad f(a,b) := \sum_{n=-\infty}^{\infty} a^{n(n+1)/2} b^{n(n-1)/2}, \qquad |ab| < 1.$$

Basic properties satisfied by $f(a,b)$ include [5, p. 34, Entry 18]

$$(2.2) \qquad f(a,b) = f(b,a),$$
$$(2.3) \qquad f(1,a) = 2f(a,a^3),$$
$$(2.4) \qquad f(-1,a) = 0,$$

and, if n is an integer,

$$(2.5) \qquad f(a,b) = a^{n(n+1)/2} b^{n(n-1)/2} f(a(ab)^n, b(ab)^{-n}).$$

The basic property (2.2) will be used many times in the sequel without comment. The function $f(a,b)$ satisfies the well-known Jacobi triple product identity [5, p. 35, Entry 19]

$$(2.6) \qquad f(a,b) = (-a;ab)_\infty (-b;ab)_\infty (ab;ab)_\infty.$$

The three most important special cases of (2.1) are

$$(2.7) \qquad \varphi(q) := f(q,q) = \sum_{n=-\infty}^{\infty} q^{n^2} = (-q;q^2)_\infty^2 (q^2;q^2)_\infty,$$

$$(2.8) \qquad \psi(q) := f(q,q^3) = \sum_{n=0}^{\infty} q^{n(n+1)/2} = \frac{(q^2;q^2)_\infty}{(q;q^2)_\infty},$$

and

$$(2.9) \quad f(-q) := f(-q,-q^2) = \sum_{n=-\infty}^{\infty} (-1)^n q^{n(3n-1)/2} = (q;q)_\infty =: q^{-1/24} \eta(\tau),$$

where $q = \exp(2\pi i\tau)$, Im $\tau > 0$, and η denotes the Dedekind eta–function. The product representations in (2.7)–(2.9) are special cases of (2.6). Also, after Ramanujan, define

$$(2.10) \qquad \chi(q) := (-q;q^2)_\infty.$$

Using (2.6) and (2.9), we can rewrite the Rogers–Ramanujan identities (1.2) in the forms

$$(2.11) \qquad G(q) = \frac{f(-q^2,-q^3)}{f(-q)} \quad \text{and} \quad H(q) = \frac{f(-q,-q^4)}{f(-q)}.$$

We shall use (2.11) many times in the remainder of the paper. A useful consequence of (2.11) in conjunction with the Jacobi triple product identity (2.6) is

$$G(q)H(q) = \frac{f(-q^5)}{f(-q)}. \tag{2.12}$$

Basic properties of the functions (2.7)–(2.10) include [**5**, pp. 39–40, Entries 24, 25(iii)]

$$\frac{f(q)}{f(-q)} = \frac{\psi(q)}{\psi(-q)} = \frac{\chi(q)}{\chi(-q)} = \sqrt{\frac{\varphi(q)}{\varphi(-q)}}, \tag{2.13}$$

$$\chi(q) = \frac{f(q)}{f(-q^2)} = \sqrt[3]{\frac{\varphi(q)}{\psi(-q)}} = \frac{\varphi(q)}{f(q)} = \frac{f(-q^2)}{\psi(-q)}, \tag{2.14}$$

$$f^3(-q^2) = \varphi(-q)\psi^2(q), \qquad \chi(q)\chi(-q) = \chi(-q^2), \tag{2.15}$$

$$\varphi(q)\varphi(-q) = \varphi^2(-q^2). \tag{2.16}$$

It is easy to deduce from (2.14) or (2.6) that

$$\psi(-q) = \chi(-q)f(-q^4) = \frac{f(-q)}{\chi(-q^2)}, \quad \chi(q)f(-q) = \varphi(-q^2). \tag{2.17}$$

We shall use the famous quintuple product identity, which, in Ramanujan's notation, takes the form (2.1) [**5**, p. 80, Entry 28(iv)]

$$\frac{f(-a^2, -a^{-2}q)}{f(-a, -a^{-1}q)} = \frac{1}{f(-q)}\left\{f(-a^3q, -a^{-3}q^2) + af(-a^{-3}q, -a^3q^2)\right\}, \tag{2.18}$$

where a is any complex number.

The function $f(a, b)$ also satisfies a useful addition formula. For each positive integer n, let

$$U_n := a^{n(n+1)/2}b^{n(n-1)/2} \quad \text{and} \quad V_n := a^{n(n-1)/2}b^{n(n+1)/2}.$$

Then [**5**, p. 48, Entry 31]

$$f(U_1, V_1) = \sum_{r=0}^{n-1} U_r f\left(\frac{U_{n+r}}{U_r}, \frac{V_{n-r}}{U_r}\right). \tag{2.19}$$

The Rogers–Ramanujan functions are intimately associated with the Rogers–Ramanujan continued fraction, defined by

$$R(q) := \frac{q^{1/5}}{1} + \frac{q}{1} + \frac{q^2}{1} + \frac{q^3}{1} + \cdots, \qquad |q| < 1, \tag{2.20}$$

which first appeared in a paper by Rogers [**31**] in 1894. Using the Rogers–Ramanujan identities (1.2), Rogers proved that

$$R(q) = q^{1/5}\frac{H(q)}{G(q)} = q^{1/5}\frac{(q;q^5)_\infty(q^4;q^5)_\infty}{(q^2;q^5)_\infty(q^3;q^5)_\infty}. \tag{2.21}$$

This was independently discovered by Ramanujan and can be found in his notebooks [**28**], [**5**, p. 79, Chap. 16, Entry 38(iii)].

CHAPTER 3

The Forty Identities

ENTRY 3.1.
$$(3.1) \qquad G^{11}(q)H(q) - q^2 G(q)H^{11}(q) = 1 + 11q G^6(q) H^6(q).$$

Entry 3.1 is one of two identities stated by Ramanujan without proof in [26], [27, p. 231]. As related in the Introduction, Ramanujan [26] claims that, "Each of these formulae is the simplest of a large class." Ramanujan's remark is interesting, because Entry 3.1 is the only identity among the forty in which powers of $G(q)$ or $H(q)$ appear. It would seem from Ramanujan's remark that he had further identities involving powers of $G(q)$ or $H(q)$, but no further identities of this sort are known. The first published proof of (3.1) is by H. B. C. Darling [16] in 1921. A second proof by Rogers [32] appeared in the same year. One year later, L. J. Mordell [24] found another proof.

By (2.21), the identity (3.1) is equivalent to a famous identity for the Rogers–Ramanujan continued fraction (2.20), namely,
$$(3.2) \qquad \frac{1}{R^5(q)} - 11 - R^5(q) = \frac{f^6(-q)}{qf^6(-q^5)}.$$

This equality was found by Watson in Ramanujan's notebooks [28] and proved by him [33] in order to establish claims about the Rogers–Ramanujan continued fraction communicated by Ramanujan in his first two letters to Hardy [33]. A different proof of (3.2) can be found in Berndt's book [5, pp. 265–267]. The identity (3.2) can also be found in an unpublished manuscript of Ramanujan first appearing in handwritten form with his lost notebook [29, pp. 135–177, 238–243]. An annotated account of Ramanujan's manuscript with considerable commentary and numerous references has been prepared by Berndt and K. Ono [7].

ENTRY 3.2.
$$(3.3) \qquad G(q)G(q^4) + qH(q)H(q^4) = \chi^2(q) = \frac{\varphi(q)}{f(-q^2)}.$$

Entry 3.2 was first proved in print by Rogers [32]; Watson [34] also found a proof. In fact, G. E. Andrews [1, p. 27] has shown that (3.3) follows from a very general identity in three variables found in Ramanujan's lost notebook.

ENTRY 3.3.
$$(3.4) \qquad G(q)G(q^4) - qH(q)H(q^4) = \frac{\varphi(q^5)}{f(-q^2)}.$$

Watson [34] gave a proof of (3.4).

ENTRY 3.4.
$$(3.5) \qquad G(q^{11})H(q) - q^2 G(q)H(q^{11}) = 1.$$

Entry 3.4 is the second identity offered by Ramanujan without proof in [**26**], [**27**, p. 231]. The first published proof was given by Rogers [**32**]. Watson [**34**] also gave a proof. R. Blecksmith, J. Brillhart, and I. Gerst [**12**] have shown that (3.5) follows from a very general theta function identity established by the authors.

Proofs of the next seven entries were first given by Rogers [**32**]. N. D. Baruah, J. Bora, and N. Saikia [**4**] and Baruah and Bora [**3**] have also found proofs of Entry 3.6.

ENTRY 3.5.
$$(3.6) \qquad G(q^{16})H(q) - q^3 G(q)H(q^{16}) = \chi(q^2).$$

ENTRY 3.6.
$$(3.7) \qquad G(q)G(q^9) + q^2 H(q)H(q^9) = \frac{f^2(-q^3)}{f(-q)f(-q^9)}.$$

ENTRY 3.7.
$$(3.8) \qquad G(q^2)G(q^3) + qH(q^2)H(q^3) = \frac{\chi(-q^3)}{\chi(-q)}.$$

ENTRY 3.8.
$$(3.9) \qquad G(q^6)H(q) - qG(q)H(q^6) = \frac{\chi(-q)}{\chi(-q^3)}.$$

ENTRY 3.9.
$$(3.10) \qquad G(q^7)H(q^2) - qG(q^2)H(q^7) = \frac{\chi(-q)}{\chi(-q^7)}.$$

ENTRY 3.10.
$$(3.11) \qquad G(q)G(q^{14}) + q^3 H(q)H(q^{14}) = \frac{\chi(-q^7)}{\chi(-q)}.$$

ENTRY 3.11.
$$(3.12) \qquad G(q^8)H(q^3) - qG(q^3)H(q^8) = \frac{\chi(-q)\chi(-q^4)}{\chi(-q^3)\chi(-q^{12})}.$$

ENTRY 3.12.
$$(3.13) \qquad G(q)G(q^{24}) + q^5 H(q)H(q^{24}) = \frac{\chi(-q^3)\chi(-q^{12})}{\chi(-q)\chi(-q^4)}.$$

ENTRY 3.13.
$$(3.14) \qquad G(q^9)H(q^4) - qG(q^4)H(q^9) = \frac{\chi(-q)\chi(-q^6)}{\chi(-q^3)\chi(-q^{18})}.$$

ENTRY 3.14.
$$(3.15) \qquad G(q^{36})H(q) - q^7 G(q)H(q^{36}) = \frac{\chi(-q^6)\chi(-q^9)}{\chi(-q^2)\chi(-q^3)}.$$

Entries 3.12–3.14 were first proved by Bressoud in his doctoral dissertation [**14**].

ENTRY 3.15.

$$(3.16) \quad G(q^3)G(q^7) + q^2 H(q^3)H(q^7) = G(q^{21})H(q) - q^4 G(q)H(q^{21})$$

$$= \frac{1}{2\sqrt{q}} \chi(q^{1/2})\chi(-q^{3/2})\chi(q^{7/2})\chi(-q^{21/2})$$

$$(3.17) \quad\quad -\frac{1}{2\sqrt{q}} \chi(-q^{1/2})\chi(q^{3/2})\chi(-q^{7/2})\chi(q^{21/2}).$$

The only known proofs of (3.16) and (3.17) are by Biagioli [**10**], who used the theory of modular forms.

ENTRY 3.16.

$$(3.18) \quad G(q^2)G(q^{13}) + q^3 H(q^2)H(q^{13}) = G(q^{26})H(q) - q^5 G(q)H(q^{26})$$

$$(3.19) \quad\quad = \sqrt{\frac{\chi(-q^{13})}{\chi(-q)}} - q\frac{\chi(-q)}{\chi(-q^{13})}.$$

The only known proof of (3.18) is by Bressoud [**14**], while Biagioli, using the theory of modular forms, has established the only known proof of (3.19). Biagioli's [**10**] formulation of (3.19) contains two misprints; the formula is also misnumbered as #17 instead of #18.

Proofs of the next four identities, (3.20)–(3.23), have been given by Bressoud [**14**].

ENTRY 3.17.

$$G(q)G(q^{19}) + q^4 H(q)H(q^{19}) = \frac{1}{4\sqrt{q}} \chi^2(q^{1/2})\chi^2(q^{19/2}) - \frac{1}{4\sqrt{q}} \chi^2(-q^{1/2})\chi^2(-q^{19/2})$$

$$(3.20) \quad\quad -\frac{q^2}{\chi^2(-q)\chi^2(-q^{19})}.$$

ENTRY 3.18.

$$G(q^{31})H(q) - q^6 G(q)H(q^{31}) = \frac{1}{2q}\chi(q)\chi(q^{31}) - \frac{1}{2q}\chi(-q)\chi(-q^{31})$$

$$(3.21) \quad\quad +\frac{q^3}{\chi(-q^2)\chi(-q^{62})}.$$

ENTRY 3.19.

$$(3.22) \quad \{G(q)G(q^{39}) + q^8 H(q)H(q^{39})\} f(-q)f(-q^{39})$$
$$= \{G(q^{13})H(q^3) - q^2 G(q^3)H(q^{13})\} f(-q^3)f(-q^{13})$$

$$(3.23) \quad = \frac{1}{2q}\left(\varphi(-q^3)\varphi(-q^{13}) - \varphi(-q)\varphi(-q^{39})\right).$$

ENTRY 3.20.

$$(3.24) \quad G(q)H(-q) + G(-q)H(q) = \frac{2}{\chi^2(-q^2)} = \frac{2\psi(q^2)}{f(-q^2)}.$$

ENTRY 3.21.

$$(3.25) \quad G(q)H(-q) - G(-q)H(q) = \frac{2q\psi(q^{10})}{f(-q^2)}.$$

Watson [**34**] constructed proofs of both (3.24) and (3.25).

ENTRY 3.22.
$$(3.26) \qquad G(-q)G(-q^4) + qH(-q)H(-q^4) = \chi(q^2).$$

ENTRY 3.23.
$$(3.27) \qquad G(-q^2)G(-q^3) + qH(-q^2)H(-q^3) = \frac{\chi(q)\chi(q^6)}{\chi(q^2)\chi(q^3)}.$$

ENTRY 3.24.
$$(3.28) \qquad G(-q^6)H(-q) - qH(-q^6)G(-q) = \frac{\chi(q^2)\chi(q^3)}{\chi(q)\chi(q^6)}.$$

Bressoud [14] established the three previous entries.

ENTRY 3.25.
$$(3.29) \qquad G(-q)G(q^9) - q^2 H(-q)H(q^9) = \frac{\chi(-q)\chi(q^9)}{\chi(-q^6)}.$$

Equality (3.29) is a corrected version of that given by Watson [29] and was first proved by Bressoud [14].

ENTRY 3.26.
$$(3.30) \quad G(q^{11})H(-q) + q^2 G(-q)H(q^{11})$$
$$= \frac{\chi(q^2)\chi(q^{22})}{\chi(-q^2)\chi(-q^{22})} - \frac{2q^3}{\chi(-q^2)\chi(-q^4)\chi(-q^{22})\chi(-q^{44})}.$$

Watson [34] established (3.30). The minus sign in front of the second expression on the right side of (3.30) is missing in Watson's list [29].

Our formulations of Entries 3.27 and 3.28 are slightly different from those of Ramanujan, who had reversed the hypotheses in each entry. In other words, he intended that the formulas for U and V be the conclusions in each case, with the pairs of equations, (3.33), (3.34) and (3.35), (3.36) being the conditions under which the formulas for U and V should hold. Watson proved Entry 3.27 under the same interpretation as we have given.

ENTRY 3.27. *Define*
$$(3.31) \qquad U := U(q) := G(q)G(q^{44}) + q^9 H(q)H(q^{44})$$
and
$$(3.32) \qquad V := V(q) := G(q^4)G(q^{11}) + q^3 H(q^4)H(q^{11}).$$
Then
$$(3.33) \qquad U^2 + qV^2 = \chi^3(q)\chi^3(q^{11})$$
and
$$(3.34) \qquad UV + q = \chi^2(q)\chi^2(q^{11}).$$

ENTRY 3.28. *Define*
$$U := G(q^{17})H(q^2) - q^3 G(q^2)H(q^{17}) \qquad \text{and} \qquad V := G(q)G(q^{34}) + q^7 H(q)H(q^{34}).$$
Then
$$(3.35) \qquad \frac{U}{V} = \frac{\chi(-q)}{\chi(-q^{17})}$$

and

$$(3.36) \qquad U^4V^4 - qU^2V^2 = \frac{\chi^3(-q^{17})}{\chi^3(-q)}\left(1 + q^2\frac{\chi^3(-q)}{\chi^3(-q^{17})}\right)^2.$$

Bressoud proved (3.35) in his thesis [**14**]. Biagioli claimed in [**10**] that he was going to prove (3.36), but a proof of (3.36) does not appear in his paper.

ENTRY 3.29.
$$\{G(q^2)G(q^{23}) + q^5H(q^2)H(q^{23})\}\{G(q^{46})H(q) - q^9G(q)H(q^{46})\}$$
$$(3.37) \qquad = \chi(-q)\chi(-q^{23}) + q + \frac{2q^2}{\chi(-q)\chi(-q^{23})}.$$

ENTRY 3.30.
$$(3.38) \qquad \frac{G(q^{19})H(q^4) - q^3G(q^4)H(q^{19})}{G(q^{76})H(-q) + q^{15}G(-q)H(q^{76})} = \frac{\chi(-q^2)}{\chi(-q^{38})}.$$

ENTRY 3.31.
$$(3.39) \qquad \frac{G(q^2)G(q^{33}) + q^7H(q^2)H(q^{33})}{G(q^{66})H(q) - q^{13}H(q^{66})G(q)} = \frac{\chi(-q^3)}{\chi(-q^{11})}.$$

ENTRY 3.32.
$$(3.40) \qquad \frac{G(q^3)G(q^{22}) + q^5H(q^3)H(q^{22})}{G(q^{11})H(q^6) - qG(q^6)H(q^{11})} = \frac{\chi(-q^{33})}{\chi(-q)}.$$

Using the theory of modular forms, Biagioli [**10**] constructed proofs of Entries 3.29–3.32. No other proofs are known.

ENTRY 3.33.
$$(3.41) \qquad \frac{G(q)G(q^{54}) + q^{11}H(q)H(q^{54})}{G(q^{27})H(q^2) - q^5G(q^2)H(q^{27})} = \frac{\chi(-q^3)\chi(-q^{27})}{\chi(-q)\chi(-q^9)}.$$

ENTRY 3.34.
$$\{G(q)G(-q^{19}) - q^4H(q)H(-q^{19})\}\{G(-q)G(q^{19}) - q^4H(-q)H(q^{19})\}$$
$$(3.42) \qquad = G(q^2)G(q^{38}) + q^8H(q^2)H(q^{38}).$$

Proofs of (3.41) and (3.42) have been found by Bressoud [**14**], who corrected a misprint in Watson's [**29**] formulation of (3.42).

ENTRY 3.35.
$$\{G(q)G(q^{94}) + q^{19}H(q)H(q^{94})\}\{G(q^{47})H(q^2) - q^9G(q^2)H(q^{47})\}$$
$$= \chi(-q)\chi(-q^{47}) + 2q^2 + \frac{2q^4}{\chi(-q)\chi(-q^{47})}$$
$$(3.43) \qquad + q\sqrt{4\chi(-q)\chi(-q^{47}) + 9q^2 + \frac{8q^4}{\chi(-q)\chi(-q^{47})}}.$$

The only known proof [**10**] of Entry 3.35 employs the theory of modular forms.

Observe that in most of the forty identities, $G(q)$ and $H(q)$ occur in the combinations,

$$(3.44) \qquad G(q^r)G(q^s) + q^{(r+s)/5}H(q^r)H(q^s), \qquad \text{when } r+s \equiv 0 \,(\text{mod}\,5),$$
$$(3.45) \qquad G(q^r)H(q^s) - q^{(r-s)/5}H(q^r)G(q^s), \qquad \text{when } r-s \equiv 0 \,(\text{mod}\,5),$$

or when one or both of q^r and q^s are replaced by $-q^r$ and $-q^s$, respectively, in either (3.44) or (3.45) above.

Ramanujan's identities are remarkable for several reasons. The Rogers–Ramanujan functions are associated with modular equations of degree 5 and q-products with base q^5. However, the "5" is missing on all the right sides of the identities, except for Entries 3.3 and 3.21. One would expect to see in such identities theta functions with arguments q^{5n}, for certain positive integers n, but such functions do not appear! We have one explanation for this phenomenon. In our heuristic "proofs" of five of the forty identities, we use the transformation formulas (6.1.1) and (6.1.2) in Lemma 6.1 for, respectively, $G(q)$ and $H(q)$. Appearing in these transformation formulas are $1/\cos(2\pi/5)$ and $1/\cos(4\pi/5)$ as multiplicative factors in (6.1.1) and (6.1.2), respectively. Also note the appearances of fifth roots of unity in the infinite products on both sides of (6.1.1) and (6.1.2). In Lemma 6.2, we show that when these products are expanded into power series, only those powers with index congruent to either 0 or 1 modulo 5 survive. In fact, Lemma 6.2 is a version of the 5-dissection of the generating function for cranks. Moreover, appearing in the coefficients when the index $n \equiv 1 \pmod 5$ are $\cos(2\pi/5)$ and $\cos(4\pi/5)$, respectively, for the products in (6.1.1) and (6.1.2). In particular, see (6.1.4) and (6.1.5), respectively. Thus, when working with the power series of $G(q^\alpha)$ and $H(q^\beta)$ in the transformed variable $\hat{q}$, considerable cancellation takes place leaving eventually only powers of 5. We then make a change of variable, replacing $\hat{q}^5$ by, say, u, and so the prominence of "5" disappears.

Next, observe that the right sides in almost all of the identities are expressed entirely in terms of the modular function χ with no other theta function appearing. We have no explanation for this phenomenon. It seems likely that the function χ played a more important role in Ramanujan's thinking than we are able to discern.

As we shall see in the proofs throughout the paper, some of the identities are amenable to general techniques established either by Watson, Rogers, or the authors. However, for those identities that are more difficult to prove (and there are many), these ideas do not appear to be useable. It was unsettling for us to find a proof of a certain identity with a great deal of effort and then discover that our ideas were inapplicable to any of the remaining identities that we sought to prove. In other words, each of the "hard" identities required an argument that seems to apply to only that identity. Thus, the authors feel that if Ramanujan did indeed have ironclad proofs for each of his identities, he had at least one key idea that all researchers to date have missed. It seems likely that the function χ played an important role in Ramanujan's primary idea(s). Each of the forty identities, in principle, can be associated with modular equations of a certain degree. It happens that for each such degree, Ramanujan recorded at least one modular equation of that degree in his notebooks [28], [5]. We conjecture that Ramanujan utilized modular equations to prove some of the forty identities in manners that we have not been able to discern.

Before embarking on the proofs, we summarize here those proofs that we have borrowed from others and those entries that we are unable to prove. The proofs of Entries 3.18 and 3.28 that we give are due to Bressoud [14]. Our proof of Entry 3.34 is a modification of his proof [14]. Our proof of Entry 3.19 begins at the same point as that of Bressoud but diverges thereafter. We give two proofs of Entry 3.12, one of which is due to Bressoud [14]. Watson's proofs of Entries 3.3, 3.21,

3.26, and 3.27 are provided. Rogers's proofs of Entries 3.9–3.11 are given. Entries 3.28 (second part), 3.29, 3.30, 3.31, and 3.35 are those we are unable to prove. (We remark that in the sequel we prove that Entries 3.31 and 3.32 are equivalent, and so it suffices to prove just one of these identities.) In Section 6, we employ asymptotic formulas for $G(q)$ and $H(q)$ to demonstrate the probable truth of these five entries.

CHAPTER 4

The Principal Ideas Behind the Proofs

In this section, we describe the main ideas behind the proofs given by Watson [34], Rogers [32], Bressoud [14], and the present authors.

We first discuss an idea of Watson [34]. In these proofs, one expresses the left sides of the identities in terms of theta functions by using (2.11). In some cases, after clearing fractions, the right side can be expressed as a product of two theta functions, say with summation indices m and n. One then tries to find a change of indices of the form

$$\alpha m + \beta n = 5M + a \quad \text{and} \quad \gamma m + \delta n = 5N + b,$$

so that the product on the right side decomposes into the requisite sum of two products of theta functions on the left side. We emphasize that this method only works when the right side is a product of two theta functions, and even then, in only some cases, does this kind of change of variables produce the desired equality. This method was probably not that used by Ramanujan, because it would seem that the identity to be proved must be explicitly known in advance.

We next present a modest generalization of Rogers's method [32]. We let p and m denote odd positive integers with $p > 1$, and let α, β, and λ be real numbers such that

(4.1) $$\alpha m^2 + \beta = \lambda p.$$

The special case when α, β, and λ are integers is given by Rogers [32]. Consider, for any real number v, the product

$$q^{p\alpha m^2 v^2} f(-q^{p\alpha+2p\alpha mv}, -q^{p\alpha-2p\alpha mv}) q^{p\beta v^2} f(-q^{p\beta+2p\beta v}, -q^{p\beta-2p\beta v})$$

(4.2) $$= \sum_{r=-\infty}^{\infty} (-1)^r q^{p\alpha(r+mv)^2} \sum_{s=-\infty}^{\infty} (-1)^s q^{p\beta(s+v)^2} = \sum_{r=-\infty}^{\infty} \sum_{s=-\infty}^{\infty} (-1)^{r+s} q^I,$$

where

$$I = p\alpha(r+mv)^2 + p\beta(s+v)^2.$$

For fixed s, write $r = ms + t$. Then, by (4.1),

$$I = p\alpha\{(s+v)m + t\}^2 + p\beta(s+v)^2$$
$$= \lambda p^2(s+v)^2 + 2p\alpha mt(s+v) + p\alpha t^2$$
$$= \lambda\left\{p(s+v) + \frac{\alpha mt}{\lambda}\right\}^2 - \frac{\alpha^2 m^2 t^2}{\lambda} + p\alpha t^2$$

(4.3) $$= \lambda\left\{p(s+v) + \frac{\alpha mt}{\lambda}\right\}^2 + \frac{\alpha\beta}{\lambda}t^2.$$

Note also that, since m is odd,

(4.4) $$(-1)^{r+s} = (-1)^t.$$

Now let

(4.5) $$S_p := \left\{\frac{1}{2p}, \frac{3}{2p}, \ldots, \frac{2p-1}{2p}\right\}.$$

Thus, using (4.2)–(4.5), we find that

(4.6)
$$\sum_{v \in S_p} q^{p\alpha m^2 v^2} f(-q^{p\alpha+2p\alpha mv}, -q^{p\alpha-2p\alpha mv}) q^{p\beta v^2} f(-q^{p\beta+2p\beta v}, -q^{p\beta-2p\beta v})$$

$$= \sum_{v \in S_p} \sum_{r=-\infty}^{\infty} (-1)^r q^{p\alpha(r+mv)^2} \sum_{s=-\infty}^{\infty} (-1)^s q^{p\beta(s+v)^2} = \sum_{k=1}^{p} \sum_{s=-\infty}^{\infty} \sum_{t=-\infty}^{\infty} (-1)^t q^I,$$

where

$$I = I(r,s,t) := \lambda \left\{ p\left(s + \frac{2k-1}{2p}\right) + \frac{\alpha m t}{\lambda}\right\}^2 + \frac{\alpha\beta}{\lambda} t^2$$

$$= \lambda \left\{ ps + k - 1 + \frac{1}{2} + \frac{\alpha m t}{\lambda}\right\}^2 + \frac{\alpha\beta}{\lambda} t^2$$

(4.7)
$$= \lambda \left\{ u + \frac{1}{2} + \frac{\alpha m t}{\lambda}\right\}^2 + \frac{\alpha\beta}{\lambda} t^2,$$

upon letting $u := ps + k - 1$. Hence, (4.6) can now be expressed as

$$\sum_{v \in S_p} q^{p\alpha m^2 v^2} f(-q^{p\alpha+2p\alpha mv}, -q^{p\alpha-2p\alpha mv}) q^{p\beta v^2} f(-q^{p\beta+2p\beta v}, -q^{p\beta-2p\beta v})$$

(4.8) $$= \sum_{u=-\infty}^{\infty} \sum_{t=-\infty}^{\infty} (-1)^t q^I,$$

with I as given in (4.7).

The strategy of Rogers is to find two sets of parameters $\{\alpha_1, \beta_1, m_1, p_1, \lambda_1\}$ and $\{\alpha_2, \beta_2, m_2, p_2, \lambda_2\}$ both giving rise to the same function on the right-hand side of (4.8). This would establish an identity between two sums of products of two theta functions each of the form (4.2). For instance, if we choose the two sets of parameters such that

(4.9) $\alpha_1\beta_1 = \alpha_2\beta_2,$ $\lambda_1 = \lambda_2,$ and $\dfrac{\alpha_1 m_1}{\lambda_1} \pm \dfrac{\alpha_2 m_2}{\lambda_2}$ is an integer,

then both sets of parameters would satisfy the formula for I in (4.7), thus giving rise to the same function on the right-hand side of (4.8).

We next show that the contributions of the terms with indices k and $p-k+1$ are identical. Applying (2.5) with $n = -m$, we find that

(4.10)
$$q^{\alpha m^2 (2k-1)^2/(4p)} f(-q^{p\alpha+\alpha m(2k-1)}, -q^{p\alpha-\alpha m(2k-1)})$$
$$= q^{\alpha m^2 (2k-1)^2/(4p) + m^2 p\alpha - m^2\alpha(2k-1)} f(-q^{p\alpha+\alpha m(2k-1)-2p\alpha m}, -q^{p\alpha-\alpha m(2k-1)+2p\alpha m})$$
$$= q^{\alpha m^2 (2p-2k+1)^2/(4p)} f(-q^{p\alpha+\alpha m(2p-2k+1)}, -q^{p\alpha-\alpha m(2p-2k+1)}),$$

where we have used the fact that p is odd. The same argument holds for the other theta function in (4.2). This establishes our claim.

Next, we show that the contribution of the term with $k = (p+1)/2$, i.e., $v = 1/2$, equals 0. Thus, we examine

$$(4.11) \qquad \sum_{r=-\infty}^{\infty} (-1)^r q^{p\alpha(r+m/2)^2} = q^{p\alpha m^2/4} f(-q^{p\alpha(1-m)}, -q^{p\alpha(1+m)}).$$

To the theta function in (4.11), we apply (2.5) with $n = (m-1)/2$. Thus, for some constant c,

$$(4.12) \qquad f(-q^{p\alpha(1-m)}, -q^{p\alpha(1+m)}) = q^c f(-1, -q^{2p\alpha}) = 0,$$

by (2.4). The same argument shows that the other theta function appearing in (4.6) also vanishes when $v = 1/2$.

Using (4.10) and (4.12) in (4.6), we deduce that, when p is odd,

$$\sum_{k=1}^{(p-1)/2} F(\alpha, \beta, m, p, \lambda, k) := \sum_{k=1}^{(p-1)/2} \sum_{r=-\infty}^{\infty} (-1)^r q^{p\alpha(r+m(2k-1)/(2p))^2}$$

$$\sum_{s=-\infty}^{\infty} (-1)^s q^{p\beta(s+(2k-1)/(2p))^2}$$

$$= \sum_{k=1}^{(p-1)/2} q^{\alpha m^2 (2k-1)^2/(4p)} f(-q^{p\alpha+\alpha m(2k-1)}, -q^{p\alpha-\alpha m(2k-1)})$$

$$\times q^{\beta(2k-1)^2/(4p)} f(-q^{p\beta+\beta(2k-1)}, -q^{p\beta-\beta(2k-1)})$$

$$(4.13) \qquad = \frac{1}{2} \sum_{u=-\infty}^{\infty} \sum_{t=-\infty}^{\infty} (-1)^t q^I,$$

where I is given in (4.7).

If p is even and if α is even, then the same argument shows that the terms with indices k and $p - k + 1$ are identical. Hence, for p even,

$$\sum_{k=1}^{p/2} F(\alpha, \beta, m, p, \lambda, k) := \sum_{k=1}^{p/2} \sum_{r=-\infty}^{\infty} (-1)^r q^{p\alpha(r+m(2k-1)/(2p))^2}$$

$$\times \sum_{s=-\infty}^{\infty} (-1)^s q^{p\beta(s+(2k-1)/(2p))^2}$$

$$= \sum_{k=1}^{p/2} q^{\alpha m^2(2k-1)^2/(4p)} f(-q^{p\alpha+\alpha m(2k-1)}, -q^{p\alpha-\alpha m(2k-1)})$$

$$\times q^{\beta(2k-1)^2/(4p)} f(-q^{p\beta+\beta(2k-1)}, -q^{p\beta-\beta(2k-1)})$$

$$(4.14) \qquad = \frac{1}{2} \sum_{u=-\infty}^{\infty} \sum_{t=-\infty}^{\infty} (-1)^t q^I,$$

where I is given in (4.7).

For later applications, we record some special cases of (4.13) and (4.14). If $p = 5$ and $m = 1$,

$$\sum_{k=1}^{2} F(\alpha, \beta, 1, 5, \lambda, k) = q^{(\alpha+\beta)/20} f(-q^{4\alpha}, -q^{6\alpha}) f(-q^{4\beta}, -q^{6\beta})$$
(4.15)
$$+ q^{9(\alpha+\beta)/20} f(-q^{2\alpha}, -q^{8\alpha}) f(-q^{2\beta}, -q^{8\beta}).$$

If $p = 5$ and $m = 3$,

$$\sum_{k=1}^{2} F(\alpha, \beta, 3, 5, \lambda, k) = q^{(9\alpha+\beta)/20} f(-q^{2\alpha}, -q^{8\alpha}) f(-q^{4\beta}, -q^{6\beta})$$
(4.16)
$$- q^{(\alpha+9\beta)/20} f(-q^{4\alpha}, -q^{6\alpha}) f(-q^{2\beta}, -q^{8\beta}),$$

where we applied (2.5) with $n = 1$. If $p = 3$ and $m = 1$,

(4.17) $$\sum_{k=1}^{1} F(\alpha, \beta, 1, 3, \lambda, k) = q^{(\alpha+\beta)/12} f(-q^{2\alpha}) f(-q^{2\beta}).$$

If $p = 2$ and $m = 1$, by (2.8),

$$\sum_{k=1}^{1} F(\alpha, \beta, 1, 2, \lambda, k) = q^{(\alpha+\beta)/8} f(-q^{\alpha}, -q^{3\alpha}) f(-q^{\beta}, -q^{3\beta})$$
(4.18)
$$= q^{(\alpha+\beta)/8} \psi(-q^{\alpha}) \psi(-q^{\beta}).$$

Rogers's ideas were extended by Bressoud [14], but we have not employed Bressoud's more general theorems in this paper. We have used Rogers's method, however, in proving further identities in Ramanujan's list.

A third approach is a method of *elimination*. Here one sets, $T(q)$, say, equal to the left side of the identity to be proved. By changes of variable, if necessary, one records two further (previously proved) identities involving $G(q)$ and $H(q)$, each involving a pair of the same Rogers–Ramanujan functions appearing in the identity to be proved. Thus, we have three equations involving the same three Rogers–Ramanujan functions, which we proceed to eliminate from the three equations. There remains then an identity involving $T(q)$ and (usually) theta functions to be proved. It must be emphasized that this method can only be applied if one can find two identities related to the one to be proved. In particular, the method cannot be utilized in those cases where Ramanujan offered only one or two identities of a given degree. The theta function identity to be verified is usually difficult, and generally one should convert it to a modular equation. Hopefully, the modular equation is a known one, in particular, one of the couple hundred that Ramanujan found, but, of course, it may not be. For completeness, we next define a modular equation.

We give the definition of a modular equation, as understood by Ramanujan. Let K, K', L, and L' denote complete elliptic integrals of the first kind associated with the moduli k, $k' := \sqrt{1-k^2}$, ℓ, and $\ell' := \sqrt{1-\ell^2}$, respectively, where $0 < k, \ell < 1$. Suppose that

(4.19) $$n \frac{K'}{K} = \frac{L'}{L}$$

for some positive rational integer n. A relation between k and ℓ induced by (4.19) is called a *modular equation of degree* n. Following Ramanujan, set

$$\alpha = k^2 \quad \text{and} \quad \beta = \ell^2.$$

We often say that β has degree n over α. If

(4.20) $$q = \exp(-\pi K'/K),$$

one of the most fundamental relations in the theory of elliptic functions is given by the formula [**5**, pp. 101–102],

(4.21) $$\varphi^2(q) = {}_2F_1(\tfrac{1}{2},\tfrac{1}{2};1;k^2) = \frac{2}{\pi}\int_0^{\pi/2}\frac{d\theta}{\sqrt{1-k^2\sin^2\theta}} =: \frac{2}{\pi}K(k).$$

The first equality in (4.21) and elementary theta function identities make it possible to write each modular equation as a theta function identity. (The second equality in (4.21) arises from expanding the integrand in a binomial series and integrating termwise.) Lastly, the multiplier m of degree n is defined by

(4.22) $$m = \frac{\varphi^2(q)}{\varphi^2(q^n)}.$$

Ramanujan derived an extensive "catalogue" of formulas [**5**, pp. 122–124] giving the "evaluations" of $f(-q)$, $\varphi(q)$, $\psi(q)$, and $\chi(q)$ at various powers of the arguments in terms of

$$z := z_1 := {}_2F_1(\tfrac{1}{2},\tfrac{1}{2};1;\alpha), \quad \alpha, \quad \text{and} \quad q.$$

If q is replaced by q^n, then the evaluations are given in terms of

$$z_n := {}_2F_1(\tfrac{1}{2},\tfrac{1}{2};1;\beta), \quad \beta, \quad \text{and} \quad q^n,$$

where β has degree n over α.

We utilize a new fourth approach in this paper in which $G(q)$ and $H(q)$ are expressed as linear combinations of G and H with arguments q^n for certain positive integers n. Watson [**34**] discovered the first pair of formulas of this sort, but used them to prove only one of the forty identities. We develop further formulas of this kind and employ them in proving over a dozen of the forty identities.

We provide here statements and proofs of the lemmas from [**9**] that we use in the sequel to establish several of Ramanujan's forty identities. Some of our proofs below *actually use some of Ramanujan's forty identities*. Indeed, some of our arguments are circular. However, in all such instances, we exhibit at least one further proof of each particular entry, which is independent of the other entries. Moreover, our arguments then show that certain pairs of entries are equivalent; for example, Entries 3.7 and 3.12 are equivalent.

We begin with Watson's lemma [**34**], Lemma 4.1. Watson's proof of (4.23) [**34**, p. 60] is based on Entries 3.2 and 3.3. Here, we provide a direct proof.

LEMMA 4.1. *With $f(-q)$ defined by* (2.9),

(4.23) $$G(q) = \frac{f(-q^8)}{f(-q^2)}\left(G(q^{16}) + qH(-q^4)\right),$$

(4.24) $$H(q) = \frac{f(-q^8)}{f(-q^2)}\left(q^3 H(q^{16}) + G(-q^4)\right).$$

Proof. Employing (2.18) with q replaced by q^{10} and a replaced by q, we find that

$$(4.25) \qquad \frac{f(-q^2,-q^8)f(-q^{10})}{f(-q,-q^9)} = f(-q^{13},-q^{17}) + qf(-q^7,-q^{23}).$$

The left hand side of (4.25), by (2.6) and (2.11), is easily seen to be equal to $f(-q^2)G(q)$, and so we conclude that

$$(4.26) \qquad f(-q^2)G(q) = f(-q^{13},-q^{17}) + qf(-q^7,-q^{23}).$$

Similarly replacing q by q^{10}, q^5, q^5, and a by $q^7, -q, -q^2$, respectively, in (2.18) and using (2.6) and (2.11), we find that

$$(4.27) \qquad f(-q^2)H(q) = f(-q^{11},-q^{19}) + q^3 f(-q,-q^{29}),$$
$$(4.28) \qquad f(-q)G(q^2) = f(q^7,q^8) - qf(q^2,q^{13}),$$

and

$$(4.29) \qquad f(-q)H(q^2) = f(q^4,q^{11}) - qf(q,q^{14}).$$

Using (2.19) twice with $n=2$, and with $U_r = (-1)^r q^{15r^2-2r}, V_r = (-1)^r q^{15r^2+2r}$ and $U_r = (-1)^r q^{15r^2-8r}, V_r = (-1)^r q^{15r^2+8r}$, respectively, we separate each term on the right side of (4.26) into its even and odd parts and so find that

$$\begin{aligned} f(-q^2)G(q) &= f(q^{56},q^{64}) - q^{13}f(q^4,q^{116}) + q\left(f(q^{44},q^{76}) - q^7 f(q^{16},q^{104})\right) \\ &= f(q^{56},q^{64}) - q^8 f(q^{16},q^{104}) + q\left(f(q^{44},q^{76}) - q^{12} f(q^4,q^{116})\right) \\ &= f(-q^8)G(q^{16}) + qf(-q^8)H(-q^4), \end{aligned}$$

where in the last step we used (4.28) and (4.27) with q replaced by q^8 and $-q^4$, respectively. This proves (4.23). The related identity (4.24) is proved in a similar way, and so we omit the details. $\square$

LEMMA 4.2. *With χ defined by (2.10),*

$$(4.30) \qquad \chi(-q)\chi(q^3)G(q) = \frac{\chi(q^6)}{\chi(-q^4)}G(-q^6) - q^5 \frac{\chi(q^2)}{\chi(-q^{12})}H(q^{24}),$$

$$(4.31) \qquad \chi(-q)\chi(q^3)H(q) = -q\frac{\chi(q^6)}{\chi(-q^4)}H(-q^6) + \frac{\chi(q^2)}{\chi(-q^{12})}G(q^{24}).$$

Proof. By two applications of Entry 3.7, the second with q replaced by $-q$, and by Entry 3.20 with q replaced by q^3,

$$(4.32)$$
$$\frac{\chi(-q^3)}{\chi(-q)}G(-q^3) - \frac{\chi(q^3)}{\chi(q)}G(q^3)$$
$$= \left(G(q^2)G(q^3) + qH(q^2)H(q^3)\right)G(-q^3) - \left(G(q^2)G(-q^3) - qH(q^2)H(-q^3)\right)G(q^3)$$
$$= qH(q^2)\left\{H(q^3)G(-q^3) + H(-q^3)G(q^3)\right\} = 2q\frac{H(q^2)}{\chi^2(-q^6)},$$

which, by (2.15), simplifies to

$$(4.33) \qquad \chi(q)\chi(-q^3)G(-q^3) - \chi(-q)\chi(q^3)G(q^3) = 2q\frac{\chi(-q^2)}{\chi^2(-q^6)}H(q^2).$$

Employing (4.23) with q replaced by $-q^3$ and q^3, respectively, in (4.33), we find that

(4.34)
$$L(q) := \chi(q)\chi(-q^3)\left\{G(q^{48}) - q^3 H(-q^{12})\right\} - \chi(-q)\chi(q^3)\left\{G(q^{48}) + q^3 H(-q^{12})\right\}$$
$$= 2q\frac{f(-q^6)\chi(-q^2)}{f(-q^{24})\chi^2(-q^6)}H(q^2) = 2q\chi(-q^2)\chi(q^6)H(q^2),$$

by (2.14) and (2.15). Collecting terms on the left side of (4.34) and using (5.7.10) below, we find that

(4.35)
$$L(q) = \left\{\chi(q)\chi(-q^3) - \chi(-q)\chi(q^3)\right\}G(q^{48})$$
$$\quad - q^3\left\{\chi(q)\chi(-q^3) + \chi(-q)\chi(q^3)\right\}H(-q^{12})$$
$$= 2q\frac{\chi(q^4)}{\chi(-q^{24})}G(q^{48}) - 2q^3\frac{\chi(q^{12})}{\chi(-q^8)}H(-q^{12}).$$

Hence, by (4.34) and (4.35),

$$2q\frac{\chi(q^4)}{\chi(-q^{24})}G(q^{48}) - 2q^3\frac{\chi(q^{12})}{\chi(-q^8)}H(-q^{12}) = 2q\chi(-q^2)\chi(q^6)H(q^2).$$

Dividing both sides by $2q$ and then replacing q^2 by q, we deduce (4.31). The companion equality (4.30) is proved in a similar way, and so we omit the details. $\square$

LEMMA 4.3. *We have*

(4.36) $\qquad \chi(q)\chi(-q^3)G(q^9) - \chi(-q)\chi(q^3)G(-q^9) = 2q\dfrac{G(q^4)}{\chi(-q^{18})}$

and

(4.37) $\qquad \chi(q)\chi(-q^3)H(q^9) + \chi(-q)\chi(q^3)H(-q^9) = 2\dfrac{H(q^4)}{\chi(-q^{18})}.$

Proof. The proofs of (4.36) and (4.37) are very similar to the proofs of (4.30) and (4.31), except that Entry 3.13 is used instead of Entry 3.20. We only prove (4.37), since the proof of (4.36) follows along the same lines.

By two applications of Entry 3.13 and one application of Entry 3.20 with q replaced by q^9,

$$\frac{\chi(q)\chi(-q^6)}{\chi(q^3)\chi(-q^{18})}H(q^9) + \frac{\chi(-q)\chi(-q^6)}{\chi(-q^3)\chi(-q^{18})}H(-q^9)$$
$$= \left\{G(-q^9)H(q^4) + qG(q^4)H(-q^9)\right\}H(q^9)$$
$$\quad + \left\{G(q^9)H(q^4) - qG(q^4)H(q^9)\right\}H(-q^9)$$
$$= H(q^4)\left\{G(-q^9)H(q^9) + G(q^9)H(-q^9)\right\} = 2\frac{H(q^4)}{\chi^2(-q^{18})}.$$

Using (2.15) above, we complete the proof of (4.37). $\square$

LEMMA 4.4. *If*

(4.38) $\qquad a(q) = \dfrac{\chi^2(q)\chi(-q^2)}{\chi(-q^6)} \qquad and \qquad b(q) = \dfrac{\chi(-q)\chi(-q^2)}{\chi(-q^3)\chi(-q^6)},$

then
$$G(q) = a(q)G(q^6) - qb(q)H(q^4), \tag{4.39}$$
$$H(q) = qa(q)H(q^6) + b(q)G(q^4). \tag{4.40}$$

First Proof of Lemma 4.4. The equality (4.39) can be rewritten in the form
$$\frac{\chi(-q^6)}{\chi(-q^2)}G(q) = \chi^2(q)G(q^6) - q\frac{\chi(-q)}{\chi(-q^3)}H(q^4). \tag{4.41}$$

When the identities for $\dfrac{\chi(-q^6)}{\chi(-q^2)}$, $\dfrac{\chi(-q)}{\chi(-q^3)}$, and $\chi^2(q)$ are substituted from (3.8), (3.9), and (3.3), respectively, it is easy to see that (4.41) is trivially satisfied. The proof of (4.40) follows along the same lines. □

Second Proof of Lemma 4.4. Define
$$B(q) := G(q) + qH(q^4) \quad \text{and} \quad qA(q) := -H(q) + G(q^4). \tag{4.42}$$

Let us also define
$$s(q) := \frac{\chi(-q^3)}{\chi(-q)}. \tag{4.43}$$

From the definition (4.42) and (3.3), we see that
$$-q^2 A(q)H(q^4) + B(q)G(q^4) = G(q)G(q^4) + qH(q)H(q^4) = \chi^2(q). \tag{4.44}$$

Similarly, by (4.42), (3.9), (3.8), and (4.43), we find that
$$qA(q)G(q^6) + qB(q)H(q^6)$$
$$= -H(q)G(q^6) + qG(q)H(q^6) + \{G(q^4)G(q^6) + q^2 H(q^4)H(q^6)\}$$
$$= -\frac{1}{s(q)} + s(q^2). \tag{4.45}$$

By (3.8) and (4.43), we solve for $B(q)$ and $qA(q)$ in (4.44) and (4.45) and find that
$$B(q) = \frac{\chi^2(q)}{s(q^2)}G(q^6) - q\frac{1}{s(q)s(q^2)}H(q^4) + qH(q^4),$$
$$qA(q) = -\frac{1}{s(q)s(q^2)}G(q^4) - q\frac{\chi^2(q)}{s(q^2)}H(q^6) + G(q^4),$$
which, by (4.42), immediately yield (4.39) and (4.40). □

Our fifth approach uses a formula of R. Blecksmith, J. Brillhart, and I. Gerst [13] providing a representation for a product of two theta functions as a sum of m products of pairs of theta functions, under certain conditions. This formula generalizes formulas of H. Schröter [5, pp. 65–72], which have been enormously useful in establishing many of Ramanujan's modular equations [5].

Define, for $\epsilon \in \{0, 1\}$ and $|ab| < 1$,
$$f_\epsilon(a, b) = \sum_{n=-\infty}^{\infty} (-1)^{\epsilon n}(ab)^{n^2/2}(a/b)^{n/2}. \tag{4.46}$$

THEOREM 4.1. *Let $a, b, c,$ and d denote positive numbers with $|ab|, |cd| < 1$. Suppose that there exist positive integers $\alpha, \beta,$ and m such that*
$$(ab)^\beta = (cd)^{\alpha(m-\alpha\beta)}. \tag{4.47}$$

Let $\epsilon_1, \epsilon_2 \in \{0,1\}$, and define $\delta_1, \delta_2 \in \{0,1\}$ by

(4.48) $\qquad \delta_1 \equiv \epsilon_1 - \alpha\epsilon_2 \pmod{2} \qquad \text{and} \qquad \delta_2 \equiv \beta\epsilon_1 + p\epsilon_2 \pmod{2}$,

respectively, where $p = m - \alpha\beta$. Then, if R denotes any complete residue system modulo m,

$$f_{\epsilon_1}(a,b)f_{\epsilon_2}(c,d) = \sum_{r \in R} (-1)^{\epsilon_2 r} c^{r(r+1)/2} d^{r(r-1)/2}$$

$$\times f_{\delta_1}\left(\frac{a(cd)^{\alpha(\alpha+1-2r)/2}}{c^\alpha}, \frac{b(cd)^{\alpha(\alpha+1+2r)/2}}{d^\alpha}\right)$$

(4.49) $\qquad \times f_{\delta_2}\left(\frac{(b/a)^{\beta/2}(cd)^{p(m+1-2r)/2}}{c^p}, \frac{(a/b)^{\beta/2}(cd)^{p(m+1+2r)/2}}{d^p}\right).$

Proof. Setting $s = k - \alpha n$, we find that

$$f_{\epsilon_1}(a,b)f_{\epsilon_2}(c,d) = \sum_{n,s=-\infty}^{\infty} (-1)^{\epsilon_1 n + \epsilon_2 s}(ab)^{n^2/2}(a/b)^{n/2}(cd)^{s^2/2}(c/d)^{s/2}$$

$$= \sum_{n,k=-\infty}^{\infty} (-1)^{\epsilon_1 n + \epsilon_2(k-\alpha n)}(ab)^{n^2/2}(a/b)^{n/2}(cd)^{(k-\alpha n)^2/2}(c/d)^{(k-\alpha n)/2}.$$

Expand into residue classes modulo m and set $k = tm + r, -\infty < t < \infty, r \in R$, to deduce that

$$f_{\epsilon_1}(a,b)f_{\epsilon_2}(c,d) = \sum_{r \in R} \sum_{n,t=-\infty}^{\infty} (-1)^{\epsilon_1 n + \epsilon_2(tm+r-\alpha n)}$$

$$\times (ab)^{n^2/2}(a/b)^{n/2}(cd)^{(tm+r-\alpha n)^2/2}(c/d)^{(tm+r-\alpha n)/2}.$$

Next, setting $n = \ell + \beta t, -\infty < \ell < \infty$, we find that

$$f_{\epsilon_1}(a,b)f_{\epsilon_2}(c,d) = \sum_{r \in R} \sum_{\ell, t=-\infty}^{\infty} (-1)^{\epsilon_1(\ell+\beta t)+\epsilon_2(tm+r-\alpha(\ell+\beta t))}$$

$$\times (ab)^{(\ell+\beta t)^2/2}(a/b)^{(\ell+\beta t)/2}(cd)^{(tm+r-\alpha(\ell+\beta t))^2/2}(c/d)^{(tm+r-\alpha(\ell+\beta t))/2}.$$

Recalling that $p = m - \alpha\beta$ and noting that $tm + r - \alpha(\ell + \beta t) = tp + r - \alpha\ell$, we find that

$$f_{\epsilon_1}(a,b)f_{\epsilon_2}(c,d) = \sum_{r \in R} \sum_{\ell, t=-\infty}^{\infty} (-1)^{\epsilon_1(\ell+\beta t)}(-1)^{\epsilon_2(tp+r-\alpha\ell)}$$

$$\times (ab)^{(\ell+\beta t)^2/2}(a/b)^{(\ell+\beta t)/2}(cd)^{(tp+r-\alpha\ell)^2/2}(c/d)^{(tp+r-\alpha\ell)/2}.$$

Now, by (4.47) and the definition $p = m - \alpha\beta$, we find that

$$(ab)^{\beta(\ell t + \beta t^2/2)}(cd)^{t^2 p^2/2 - tp\alpha\ell} = (cd)^{\alpha p(\ell t + \beta t^2/2)}(cd)^{t^2 p^2/2 - tp\alpha\ell}$$

$$= (cd)^{t^2 p(\alpha\beta + p)/2}$$

$$= (cd)^{t^2 pm/2}.$$

Hence, recalling the definitions of δ_1 and δ_2 from (4.48) and the definition of $f_\epsilon(a,b)$ from (4.46), we find that

$$f_{\epsilon_1}(a,b) f_{\epsilon_2}(c,d) = \sum_{r \in R} \sum_{\ell, t = -\infty}^{\infty} (-1)^{\delta_1 \ell} (-1)^{\delta_2 t} (-1)^{\epsilon_2 r} (cd)^{r^2/2} (c/d)^{r/2}$$

$$\times \left(ab(cd)^{\alpha^2} \right)^{\ell^2/2} \left(\frac{a}{b} \left(\frac{c}{d} \right)^{-\alpha} (cd)^{-2r\alpha} \right)^{\ell/2} ((cd)^{mp})^{t^2/2} \left(\left(\frac{a}{b} \right)^\beta \left(\frac{c}{d} \right)^p (cd)^{2pr} \right)^{t/2}$$

$$= \sum_{r \in R} (-1)^{\epsilon_2 r} c^{r(r+1)/2} d^{r(r-1)/2} f_{\delta_1} \left(\frac{a(cd)^{\alpha(\alpha+1-2r)/2}}{c^\alpha}, \frac{b(cd)^{\alpha(\alpha+1+2r)/2}}{d^\alpha} \right)$$

$$\times f_{\delta_2} \left(\frac{(b/a)^{\beta/2} (cd)^{p(m+1-2r)/2}}{c^p}, \frac{(a/b)^{\beta/2} (cd)^{p(m+1+2r)/2}}{d^p} \right),$$

after some elementary algebra and elementary manipulation. $\square$

CHAPTER 5

Proofs of 35 of the 40 Entries

5.1. Proof of Entry 3.1

We begin by proving the following identity from Chapter 19 of Ramanujan's second notebook [28], [5, p. 80, Entry 38(iv); p. 262, Entry 10(iii)]. Our proof is taken from [5, pp. 81–82].

LEMMA 5.1. *We have*

(5.1.1) $\quad f^2(-q^2, -q^3) - q^{2/5} f^2(-q, -q^4) = f(-q) \left\{ f(-q^{1/5}) + q^{1/5} f(-q^5) \right\}.$

Proof. Apply (2.19) with $a = -q$, $b = -q^2$, and $n = 5$. Then

$$U_n = (-1)^n q^{n(3n-1)/2} \quad \text{and} \quad V_n = (-1)^n q^{n(3n+1)/2}.$$

Thus, by (2.9) and (2.19),

$$f(-q) = f(-q, -q^2) = f(-q^{35}, -q^{40}) - qf(-q^{50}, -q^{25}) + q^5 f(-q^{65}, -q^{10})$$
$$- q^{12} f(-q^{80}, -q^{-5}) + q^{22} f(-q^{95}, -q^{-20})$$
$$= -qf(-q^{25}) + \{f(-q^{35}, -q^{40}) + q^5 f(-q^{10}, -q^{65})\}$$
(5.1.2) $\quad - q^2 \{f(-q^{20}, -q^{55}) + q^{10} f(-q^{-5}, -q^{80})\},$

where we applied (2.5). We now invoke the quintuple product identity (2.18) twice, with q replaced by q^{25} and $a = q^5, q^{10}$, respectively. We therefore find that (5.1.2) can be written as

(5.1.3) $\quad f(-q) + qf(-q^{25}) = f(-q^{25}) \left\{ \dfrac{f(-q^{15}, -q^{10})}{f(-q^{20}, -q^5)} - q^2 \dfrac{f(-q^5, -q^{20})}{f(-q^{15}, -q^{10})} \right\}.$

By (2.6),

(5.1.4) $\quad f(-q, -q^4) f(-q^2, -q^3) = f(-q) f(-q^5).$

Multiplying both sides of (5.1.3) by $f(-q)$, but with q replaced by $q^{1/5}$, and using (5.1.4), we deduce that

$$f(-q) \left\{ f(-q^{1/5}) + q^{1/5} f(-q^5) \right\}$$
$$= f(-q) f(-q^5) \left\{ \dfrac{f(-q^2, -q^3)}{f(-q, -q^4)} - q^{2/5} \dfrac{f(-q, -q^4)}{f(-q^2, -q^3)} \right\}$$
$$= f^2(-q^2, -q^3) - q^{2/5} f^2(-q, -q^4),$$

which completes the proof. $\square$

Proof of Entry 3.1. Replace $q^{1/5}$ by $q^{1/5}\zeta$ in (5.1.1), where ζ is a fifth root of unity, and multiply all five identities together. We then find that

$$(5.1.5) \quad f^5(-q) \prod_\zeta f(-q^{1/5}\zeta)$$
$$= \prod_\zeta \left\{ f^2(-q^2, -q^3) - q^{2/5}\zeta^2 f^2(-q, -q^4) - q^{1/5}\zeta f(-q)f(-q^5) \right\}.$$

Multiplying out the products on each side of (5.1.5), we find that

$$\frac{f^{11}(-q)}{f(-q^5)} = f^{10}(-q^2, -q^3) - q^2 f^{10}(-q, -q^4) - q f^5(-q) f^5(-q^5)$$
$$- 5q f^2(-q^2, -q^3) f^2(-q, -q^4) f^3(-q) f^3(-q^5)$$
$$(5.1.6) \qquad - 5q f^4(-q^2, -q^3) f^4(-q, -q^4) f(-q) f(-q^5).$$

By (5.1.4), (5.1.6) simplifies to

$$\frac{f^{11}(-q)}{f(-q^5)} = f^{10}(-q^2, -q^3) - q^2 f^{10}(-q, -q^4) - q f^5(-q) f^5(-q^5)$$
$$- 5q f^5(-q) f^5(-q^5) - 5q f^5(-q) f^5(-q^5)$$
$$(5.1.7) \qquad = f^{10}(-q^2, -q^3) - q^2 f^{10}(-q, -q^4) - 11q f^5(-q) f^5(-q^5).$$

Multiplying both sides of (5.1.7) by $f(-q^5)/f^{11}(-q)$, using (5.1.4), and lastly employing (2.11), we conclude that

$$1 = \frac{f^{11}(-q^2, -q^3) f(-q, -q^4)}{f^{12}(-q)} - q^2 \frac{f^{11}(-q, -q^4) f(-q^2, -q^3)}{f^{12}(-q)}$$
$$- 11q \frac{f^6(-q, -q^4) f^6(-q^2, -q^3)}{f^{12}(-q)}$$
$$= G^{11}(q) H(q) - q^2 H^{11}(q) G(q) - 11q G^6(q) H^6(q),$$

which completes the proof of Entry 3.1. □

5.2. Proofs of Entry 3.2

Proof. Using (4.23) and (4.24) in (3.6), we find that

$$\chi(q^2) = G(q^{16}) H(q) - q^3 H(q^{16}) G(q)$$
$$= \frac{f(-q^8)}{f(-q^2)} \left\{ G(q^{16}) \left(q^3 H(q^{16}) + G(-q^4) \right) - q^3 H(q^{16}) \left(G(q^{16}) + q H(-q^4) \right) \right\}$$
$$= \frac{f(-q^8)}{f(-q^2)} \left\{ G(-q^4) G(q^{16}) - q^4 H(-q^4) H(q^{16}) \right\}.$$

Therefore, by (2.14) and (2.15), we deduce that

$$G(-q^4) G(q^{16}) - q^4 H(-q^4) H(q^{16}) = \frac{f(-q^2)}{f(-q^8)} \chi(q^2)$$
$$= \frac{\chi(-q^2) f(-q^4) \chi(q^2)}{f(-q^4)/\chi(-q^4)} = \chi^2(-q^4),$$

which is Entry 3.2 with q replaced by $-q^4$. □

In his lost notebook [**29**, p. 27], Ramanujan offers the beautiful identity

$$\sum_{n=0}^{\infty}\frac{a^n b^n q^{n^2}}{(q^4;q^4)_n}\sum_{n=0}^{\infty}\frac{a^{-2n}q^{4n^2}}{(bq^4;q^4)_n}+\sum_{n=0}^{\infty}\frac{a^n b^n q^{(n+1)^2}}{(q^4;q^4)_n}\sum_{n=0}^{\infty}\frac{a^{-2n-1}q^{4n^2+4n}}{(bq^4;q^4)_n}$$

(5.2.1)
$$=\frac{f(aq,q/a)}{(bq^4;q^4)_\infty}-(1-b)\sum_{n=0}^{\infty}a^{n+1}q^{(n+1)^2}\sum_{j=0}^{n}\frac{b^j}{(q^4;q^4)_j}.$$

If we set $a = b = 1$ in (5.2.1) and multiply both sides by $(-q^2;q^2)_\infty$, we see that (5.2.1) reduces to

(5.2.2) $(-q^2;q^2)_\infty \sum_{n=0}^{\infty}\frac{q^{n^2}}{(q^4;q^4)_n}G(q^4)+(-q^2;q^2)_\infty\sum_{n=0}^{\infty}\frac{q^{(n+1)^2}}{(q^4;q^4)_n}H(q^4)=\frac{\varphi(q)}{f(-q^2)}.$

However, Rogers [**31**] proved that

$$(-q^2;q^2)_\infty \sum_{n=0}^{\infty}\frac{q^{n^2}}{(q^4;q^4)_n}=G(q)$$

and

$$(-q^2;q^2)_\infty\sum_{n=0}^{\infty}\frac{q^{n^2+2n}}{(q^4;q^4)_n}=H(q),$$

and so (5.2.2) reduces to (3.3). A proof of (5.2.1) has been given by G. E. Andrews [**1**, pp. 28–33].

5.3. Proof of Entry 3.3

Entry 3.3 follows from combining Entry 3.2 with the following lemma.

LEMMA 5.2. *We have*

(5.3.1) $$\varphi^2(q)-\varphi^2(q^5)=4qf^2(-q^{10})\frac{\chi(q)}{\chi(q^5)}.$$

Proof. By Entry 10(iv) in Chapter 19 of Ramanujan's second notebook [**28**], [**5**, p. 262] and the Jacobi triple product identity (2.6),

$$\varphi^2(q)-\varphi^2(q^5)=4qf(q,q^9)f(q^3,q^7)$$
$$=4q(-q;q^{10})_\infty(-q^9;q^{10})_\infty(-q^3;q^{10})_\infty(-q^7;q^{10})_\infty(q^{10};q^{10})_\infty^2$$
$$=4qf^2(-q^{10})\frac{(-q;q^2)_\infty}{(-q^5;q^{10})_\infty}$$
$$=4qf^2(-q^{10})\frac{\chi(q)}{\chi(q^5)},$$

and the proof is complete. □

The identity (5.3.1) is an analogue of

$$\varphi^2(q)-5\varphi^2(q^5)=-4f^2(-q^2)\frac{\chi(q^5)}{\chi(q)},$$

which is found in Ramanujan's lost notebook [**29**] and was first proved by S.-Y. Kang [**20**, Thm. 2.2(i)].

Proof of Entry 3.3. The proof which follows is due to Watson [34]. From Entry 3.2, (2.11), and (5.1.4), we find that

$$\{G(q)G(q^4) - qH(q)H(q^4)\}^2 = \{G(q)G(q^4) + qH(q)H(q^4)\}^2$$
$$- 4qG(q)H(q)G(q^4)H(q^4)$$
(5.3.2)
$$= \frac{\varphi^2(q)}{f^2(-q^2)} - 4q\frac{f(-q^5)f(-q^{20})}{f(-q)f(-q^4)}.$$

A straightforward calculation shows that

(5.3.3) $$\chi(q) = \frac{f^2(-q^2)}{f(-q)f(-q^4)}.$$

Using (5.3.3) twice, we find that (5.3.2) can be written in the form

$$\{G(q)G(q^4) - qH(q)H(q^4)\}^2 = \frac{\varphi^2(q) - 4qf^2(-q^{10})\dfrac{\chi(q)}{\chi(q^5)}}{f^2(-q^2)} = \frac{\varphi^2(q^5)}{f^2(-q^2)},$$

where we applied Lemma 5.2. Taking the square root of both sides above, we complete the proof. □

5.4. Proof of Entry 3.4

By employing (2.11), we easily find that the proposed identity is equivalent to the identity

(5.4.1) $f(-q,-q^4)f(-q^{22},-q^{33}) - q^2 f(-q^2,-q^3)f(-q^{11},-q^{44}) = f(-q)f(-q^{11}).$

To prove (5.4.1), we apply the ideas of Rogers, in particular, (4.13) with the two sets of parameters $\alpha_1 = 1, \beta_1 = 11, m_1 = 3, p_1 = 5, \lambda_1 = 4$ and $\alpha_2 = 1, \beta_2 = 11, m_2 = 1, p_2 = 3, \lambda_2 = 4$. The requisite conditions (4.9) are readily seen to be satisfied. Using (4.16) and (4.17), we derive the identity

$$f(-q^2,-q^8)f(-q^{44},-q^{66}) - q^4 f(-q^4,-q^6)f(-q^{22},-q^{88}) = f(-q^2)f(-q^{22}),$$

which is the same as (5.4.1), but with q replaced by q^2.

5.5. Proof of Entry 3.5

The proof of Entry 3.5 is very similar to that for Entry 3.22 below. In fact, we reduce the desired equality to the same new modular equation (5.21.11) of degree 5. Remarkably, Ramanujan derived 27 modular equations of degree 5, although several are "reciprocals" of others [5, pp. 280–282, Entry 13]. In Ramanujan's terminology, let β have degree 5 over α.

LEMMA 5.3. *If β has degree 5 over α, then*

(5.5.1) $$(1-\beta)^{1/4} - (1-\alpha)^{1/4} = 2^{2/3}(\alpha\beta)^{1/6}\{(1-\alpha)(1-\beta)\}^{1/24}.$$

Proof. Let

$$m = \frac{\varphi^2(q)}{\varphi^2(q^5)}$$

denote the *multiplier* of degree 5. As in [5, p. 284, eq. (13.3)], define

(5.5.2) $$\rho := \sqrt{m^3 - 2m^2 + 5m}.$$

We shall need the following parameterizations for certain algebraic functions of α and β, namely [5, pp. 285–286, eqs. (13.8), (13.10), (13.11)],

$$\text{(5.5.3)} \qquad \{16\alpha\beta(1-\alpha)(1-\beta)\}^{1/6} = \frac{(m-1)(5-m)}{4m},$$

$$\text{(5.5.4)} \qquad \{(1-\alpha)^3(1-\beta)\}^{1/8} = \frac{\rho - 3m + 5}{4m},$$

and

$$\text{(5.5.5)} \qquad \{(1-\alpha)(1-\beta)^3\}^{1/8} = \frac{\rho - m^2 + 3m}{4m},$$

where ρ is defined by (5.5.2). Using (5.5.3)–(5.5.5), we find that

$$\frac{(1-\beta)^{1/4} - (1-\alpha)^{1/4}}{2^{2/3}(\alpha\beta)^{1/6}\{(1-\alpha)(1-\beta)\}^{1/24}} = \frac{\{(1-\alpha)(1-\beta)^3\}^{1/8} - \{(1-\alpha)^3(1-\beta)\}^{1/8}}{2^{2/3}\{\alpha\beta(1-\alpha)(1-\beta)\}^{1/6}}$$

$$= \frac{(\rho - m^2 + 3m) - (\rho - 3m + 5)}{(m-1)(5-m)}$$

$$= \frac{m^2 - 6m + 5}{(m-1)(m-5)} = 1,$$

which completes the proof. $\square$

We begin the proof of Entry 3.5 with the system of equations

$$G(q^{16})H(q) - q^3 G(q)H(q^{16}) =: T(q),$$

$$G(q^{16})G(q^4) + q^4 H(q^4)H(q^{16}) = \frac{\varphi(q^4)}{f(-q^8)},$$

$$G(q^{16})G(q^4) - q^4 H(q^4)H(q^{16}) = \frac{\varphi(q^{20})}{f(-q^8)}.$$

From the first equation above, we see that our task is to prove that $T(q) = \chi(q^2)$. The second and third equations are simply (3.3) and (3.4), respectively, but with q replaced by q^4. Regarding this system in the variables $G(q^{16})$, $q^3 H(q^{16})$, and -1, we see that

$$\text{(5.5.6)} \qquad \begin{vmatrix} H(q) & -G(q) & T(q) \\ G(q^4) & qH(q^4) & \dfrac{\varphi(q^4)}{f(-q^8)} \\ G(q^4) & -qH(q^4) & \dfrac{\varphi(q^{20})}{f(-q^8)} \end{vmatrix} = 0.$$

Expanding the determinant in (5.5.6) along the last column, we find that

$$\text{(5.5.7)} \qquad \begin{aligned} -2qG(q^4)H(q^4)T(q) &- \frac{\varphi(q^4)}{f(-q^8)}\{G(q)G(q^4) - qH(q)H(q^4)\} \\ &+ \frac{\varphi(q^{20})}{f(-q^8)}\{G(q)G(q^4) + qH(q)H(q^4)\} = 0. \end{aligned}$$

Using (2.12), (3.3), (3.4), and

$$\text{(5.5.8)} \qquad \frac{f(-q^4)}{f(-q^8)} = \chi(-q^4) = \chi(-q^2)\chi(q^2),$$

which arises from (2.15), we can write (5.5.7) in the form

$$\tag{5.5.9} -2q\frac{f(-q^{20})}{f(-q^4)}T(q) - \frac{\varphi(q^4)}{f(-q^8)}\frac{\varphi(q^5)}{f(-q^2)} + \frac{\varphi(q^{20})}{f(-q^8)}\frac{\varphi(q)}{f(-q^2)} = 0.$$

Rearranging (5.5.9) while using (5.5.8), we find that

$$\tag{5.5.10} 2qT(q) = \frac{\chi(-q^2)\chi(q^2)}{f(-q^2)f(-q^{20})}\left\{\varphi(q)\varphi(q^{20}) - \varphi(q^5)\varphi(q^4)\right\}.$$

Recall the representations [**5**, pp. 122–124, Entries 10(i), (v), (ii), 11(v), 12(iii), (iv), (vii)]

$$\tag{5.5.11} \varphi(q) = \sqrt{z_1}, \qquad \varphi(q^4) = \tfrac{1}{2}\sqrt{z_1}\left\{1 + (1-\alpha)^{1/4}\right\},$$

$$\tag{5.5.12} \varphi(-q) = \sqrt{z_1}(1-\alpha)^{1/4}, \qquad \psi(q^8) = \frac{1}{4q}\sqrt{z_1}\left\{1 - (1-\alpha)^{1/4}\right\},$$

$$\tag{5.5.13} f(-q^2) = \sqrt{z_1}2^{-1/3}\left(\frac{\alpha(1-\alpha)}{q}\right)^{1/12}, \qquad f(-q^4) = \sqrt{z_1}2^{-2/3}\left(\frac{(1-\alpha)\alpha^4}{q^4}\right)^{1/24},$$

and

$$\tag{5.5.14} \chi(-q^2) = 2^{1/3}\left(\frac{(1-\alpha)q^2}{\alpha^2}\right)^{1/24},$$

where

$$z_n := \varphi(q^n).$$

Recall from the theory of modular equations that, if n is the degree of the modular equation, then (5.5.13) also holds with z_1, q, and α replaced by z_n, q^n, and β, respectively, where β has degree n over α. Hence, from (5.5.13) and (5.5.14), we find that, after simplification,

$$\tag{5.5.15} \frac{\chi(-q^2)}{f(-q^2)f(-q^{20})} = \frac{2^{4/3}q}{\sqrt{z_1 z_5}(\alpha\beta)^{1/6}\left\{(1-\alpha)(1-\beta)\right\}^{1/24}}.$$

Employing (5.5.15) and (5.5.11) in (5.5.10), we deduce that

$$2qT(q) = \frac{\chi(q^2)2^{4/3}q}{\sqrt{z_1 z_5}(\alpha\beta)^{1/6}\left\{(1-\alpha)(1-\beta)\right\}^{1/24}}$$
$$\times \sqrt{z_1 z_5}\left\{\tfrac{1}{2}\left\{1 + (1-\beta)^{1/4}\right\} - \tfrac{1}{2}\left\{1 + (1-\alpha)^{1/4}\right\}\right\}$$
$$= \frac{\chi(q^2)2^{1/3}q\left\{(1-\beta)^{1/4} - (1-\alpha)^{1/4}\right\}}{(\alpha\beta)^{1/6}\left\{(1-\alpha)(1-\beta)\right\}^{1/24}}$$
$$\tag{5.5.16} = 2q\chi(q^2),$$

by Lemma 5.3. Equation (5.5.16) is trivially equivalent to (3.6), and so the proof is complete.

5.6. Proofs of Entry 3.6

First Proof of Entry 3.6. By using (2.11), we find that in order to prove Entry 3.6, it suffices to prove that

(5.6.1) $\quad f(-q^2, -q^3)f(-q^{18}, -q^{27}) + q^2 f(-q, -q^4)f(-q^9, -q^{36}) = f^2(-q^3).$

We apply (4.13) with $\alpha_1 = 1, \beta_1 = 9, m_1 = 1, p_1 = 5, \lambda_1 = 2$ and with $\alpha_2 = 3, \beta_2 = 3, m_2 = 1, p_2 = 3, \lambda_2 = 2$. We easily check that these two sets of parameters satisfy conditions (4.9). From (4.13) and (4.15), we then deduce the identity

$$f(-q^4, -q^6)f(-q^{36}, -q^{54}) + q^4 f(-q^2, -q^8)f(-q^{18}, -q^{72}) = f^2(-q^6),$$

which is precisely (5.6.1), but with q replaced by q^2. This then completes the proof of Entry 3.6. □

Second Proof of Entry 3.6. We rewrite (5.6.1) in the form

$$\sum_{\substack{m=-\infty \\ m \equiv 0 \,(\mathrm{mod}\, 3)}}^{\infty} \sum_{n=-\infty}^{\infty} (-1)^{m+n} q^{(5m^2 - 3m + 5n^2 - n)/2}$$

$$+ q^2 \sum_{\substack{m=-\infty \\ m \equiv 0 \,(\mathrm{mod}\, 3)}}^{\infty} \sum_{n=-\infty}^{\infty} (-1)^{m+n} q^{(5m^2 - 9m + 5n^2 - 3n)/2}$$

(5.6.2) $\quad = \sum_{\substack{m,n=-\infty \\ m \equiv 0, n \equiv 0 \,(\mathrm{mod}\, 3)}}^{\infty} (-1)^{m+n} q^{(m^2 - m + n^2 - n)/2} =: F(q).$

Now, for $(a, b) \in \{0, \pm 1, \pm 2\}$, set

$$2m + n = 5M + a \quad \text{and} \quad -m + 2n = 5N + b.$$

Hence,

$$m = 2M - N + (2a - b)/5 \quad \text{and} \quad n = M + 2N + (a + 2b)/5,$$

where the parameters a and b are given in the first table below. The corresponding values of m and n are given in the table which follows.

a	0	1	-1	2	-2
b	0	2	-2	-1	1
m	$2M - N$	$2M - N$	$2M - N$	$2M - N + 1$	$2M - N - 1$
n	$M + 2N$	$M + 2N + 1$	$M + 2N - 1$	$M + 2N$	$M + 2N$

Recalling that $m, n \equiv 0 \,(\mathrm{mod}\, 3)$, for the five cases in the table above, we find that, respectively,

$$M \equiv N \equiv 0 \,(\mathrm{mod}\, 3), \quad M \equiv 1, N \equiv -1 \,(\mathrm{mod}\, 3), \quad M \equiv -1, N \equiv 1 \,(\mathrm{mod}\, 3),$$
$$M \equiv N \equiv -1 \,(\mathrm{mod}\, 3), \quad M \equiv N \equiv 1 \,(\mathrm{mod}\, 3).$$

Calculating the corresponding values of $m^2 + n^2 - m - n$, we find that

$$F(q) = \sum_{\substack{M,N=-\infty \\ M \equiv 0, N \equiv 0 \,(\mathrm{mod}\, 3)}}^{\infty} (-1)^{M+N} q^{(5M^2 - 3M + 5N^2 - N)/2}$$

$$+ \sum_{\substack{M,N=-\infty \\ M\equiv 1, N\equiv -1 \pmod 3}}^{\infty} (-1)^{M+N+1} q^{(5M^2-M+5N^2+3N)/2}$$

$$+ \sum_{\substack{M,N=-\infty \\ M\equiv -1, N\equiv 1 \pmod 3}}^{\infty} (-1)^{M+N+1} q^{(5M^2-5M+5N^2-5N+2)/2}$$

$$+ \sum_{\substack{M,N=-\infty \\ M\equiv -1, N\equiv -1 \pmod 3}}^{\infty} (-1)^{M+N+1} q^{(5M^2+M+5N^2-3N)/2}$$

$$+ \sum_{\substack{M,N=-\infty \\ M\equiv 1, N\equiv 1 \pmod 3}}^{\infty} (-1)^{M+N-1} q^{(5M^2-7M+5N^2+N+2)/2}$$

(5.6.3) $\qquad =: S_1 + S_2 + S_3 + S_4 + S_5.$

First, setting $M = 3m - 1$, we find that

$$\sum_{\substack{M=-\infty \\ M\equiv -1 \pmod 3}}^{\infty} (-1)^M q^{5(M^2-M)/2} = -\sum_{m=-\infty}^{\infty} (-1)^m q^{45m(m-1)/2} = -f(-1,-q^{45}) = 0,$$

by (2.4). Hence,

(5.6.4) $\qquad\qquad\qquad\qquad S_3 = 0.$

Replacing M by $M + 1$, and then changing the signs of M and N, we readily find that

$$S_5 = \sum_{\substack{M,N=-\infty \\ M\equiv 0, N\equiv -1 \pmod 3}}^{\infty} (-1)^{M+N} q^{(5M^2+5N^2-3M-N)/2}.$$

Changing the signs of M and N and then replacing M by $M - 1$, we deduce that

$$S_2 = q^2 \sum_{\substack{M,N=-\infty \\ M\equiv 0, N\equiv 1 \pmod 3}}^{\infty} (-1)^{M+N} q^{(5M^2+5N^2-9M-3N)/2}.$$

Replacing M by $M - 1$, we easily see that

$$S_4 = q^2 \sum_{\substack{M,N=-\infty \\ M\equiv 0, N\equiv -1 \pmod 3}}^{\infty} (-1)^{M+N} q^{(5M^2+5N^2-9M-3N)/2}.$$

Hence,

(5.6.5) $\qquad S_1 + S_5 = \sum_{\substack{M,N=-\infty \\ M\equiv 0, N\equiv 0,-1 \pmod 3}}^{\infty} (-1)^{M+N} q^{(5M^2+5N^2-3M-N)/2}$

and

(5.6.6) $\qquad S_2 + S_4 = q^2 \sum_{\substack{M,N=-\infty \\ M\equiv 0, N\equiv \pm 1 \pmod 3}}^{\infty} (-1)^{M+N} q^{(5M^2+5N^2-9M-3N)/2}.$

Substituting (5.6.4)–(5.6.6) in (5.6.3) and comparing this with (5.6.2), we see that in order prove (5.6.2), we need to show that

$$\sum_{\substack{m,n=-\infty \\ m\equiv 0, n\equiv 1 \,(\text{mod } 3)}}^{\infty} (-1)^{m+n} q^{(5m^2+5n^2-3m-n)/2}$$

(5.6.7)
$$+ q^2 \sum_{\substack{m,n=-\infty \\ m,n\equiv 0 \,(\text{mod } 3)}}^{\infty} (-1)^{m+n} q^{(5m^2+5n^2-9m-3n)/2} = 0.$$

If we set $m = 3k$, $n = 3l+1$ and $m = -3l$, $n = 3k$ in the first and second sums of (5.6.7), respectively, we easily deduce (5.6.7), and so the proof is complete. □

Entry 3.6 is a natural companion to Entry 3.13; in Section 5.13, a third proof of Entry 3.6 will be concomitantly given with a proof of Entry 3.13.

5.7. Proofs of Entry 3.7

First Proof of Entry 3.7. Using (2.11), we can write (3.8) in the alternative form
(5.7.1)
$$f(-q^4,-q^6)f(-q^6,-q^9) + qf(-q^2,-q^8)f(-q^3,-q^{12}) = f(-q^2)f(-q^3)\frac{\chi(-q^3)}{\chi(-q)}.$$

Using

(5.7.2) $$\chi(-q^3) = \frac{\varphi(-q^3)}{f(-q^3)} \quad \text{and} \quad \chi(-q) = \frac{f(-q^2)}{\psi(q)}$$

from (2.14), we rewrite (5.7.1) as

(5.7.3) $$f(-q^4,-q^6)f(-q^6,-q^9) + qf(-q^2,-q^8)f(-q^3,-q^{12}) = \psi(q)\varphi(-q^3).$$

For a and b in the set $\{0, \pm 1, \pm 2\}$, let

$$m + 3n = 5M + a \quad \text{and} \quad m - 2n = 5N + b,$$

from which it follows that

$$n = M - N + (a-b)/5 \quad \text{and} \quad m = 2M + 3N + (2a+3b)/5.$$

It follows easily that $a = b$, and so $m = 2M + 3N + a$ and $n = M - N$, where $-2 \leq a \leq 2$. Thus, there is a one-to-one correspondence between the set of all pairs of integers (m, n), $-\infty < m, n < \infty$, and triples of integers (M, N, a), $-\infty < M, N < \infty$, $-2 \leq a \leq 2$. From the definitions (2.8) and (2.7) of $\psi(q)$ and $\varphi(-q^3)$, the indicated changes of indices of summation, and (2.4),

$$2\psi(q)\varphi(-q^3) = \sum_{m,n=-\infty}^{\infty} (-1)^n q^{m(m+1)/2+3n^2}$$

$$= \sum_{a=-2}^{2} q^{a(a+1)/2} \sum_{M=-\infty}^{\infty} (-1)^M q^{5M^2+(1+2a)M}$$

$$\times \sum_{N=-\infty}^{\infty} (-1)^N q^{15N^2/2+3N/2+3aN}$$

$$= \sum_{a=-2}^{2} q^{a(a+1)/2} f(-q^{4-2a}, -q^{6+2a}) f(-q^{6-3a}, -q^{9+3a})$$

$$= 2f(-q^4, -q^6)f(-q^6, -q^9) + 2qf(-q^2, -q^8)f(-q^3, -q^{12})$$
$$+ q^3 f(-1, -q^{10}) f(-1, -q^{15})$$
$$= 2f(-q^4, -q^6)f(-q^6, -q^9) + 2qf(-q^2, -q^8)f(-q^3, -q^{12}),$$

which is (5.7.3). So we complete our proof. □

Second Proof of Entry 3.7. Using (4.23) and (4.24) in (3.13), we arrive at

$$\frac{\chi(-q^3)\chi(-q^{12})}{\chi(-q)\chi(-q^4)} = G(q)G(q^{24}) + q^5 H(q)H(q^{24})$$

$$= \frac{f(-q^8)}{f(-q^2)} \{G(q^{24})\left(G(q^{16}) + qH(-q^4)\right)$$
$$+ q^5 H(q^{24})\left(q^3 H(q^{16}) + G(-q^4)\right)\}$$

$$= \frac{f(-q^8)}{f(-q^2)} \{G(q^{16})G(q^{24}) + q^8 H(q^{16})H(q^{24})$$

(5.7.4) $$+ q\left(H(-q^4)G(q^{24}) + q^4 G(-q^4)H(q^{24})\right)\}.$$

By several applications of (2.14), we deduce from (5.7.4) that

(5.7.5)

$$G(q^{16})G(q^{24}) + q^8 H(q^{16})H(q^{24}) + q\left(H(-q^4)G(q^{24}) + q^4 G(-q^4)H(q^{24})\right)$$
$$= \frac{\chi(-q^3)\chi(-q^{12})f(-q^2)}{\chi(-q)\chi(-q^4)f(-q^8)} = \chi(q)\chi(-q^3)\chi(-q^{12}).$$

Therefore, it suffices to find the even and the odd parts of $\chi(q)\chi(-q^3)$. By (2.6), (2.8), and (2.17),

$$f(-q, -q^5) = (q; q^6)_\infty (q^5; q^6)_\infty (q^6; q^6)_\infty$$

$$= \frac{(q; q^2)_\infty}{(q^3; q^6)_\infty}(q^6; q^6)_\infty$$

(5.7.6) $$= \chi(-q)\psi(q^3) = \chi(-q)\chi(q^3)f(-q^{12}).$$

Employing (2.19) with $n = 2$, $a = q$, and $b = q^5$, we also find that

(5.7.7) $$f(q, q^5) = f(q^8, q^{16}) + qf(q^4, q^{20}).$$

It is also easily verified that (see [**5**, p. 350, eq. (2.3)])

(5.7.8) $$f(q, q^2) = \frac{\varphi(-q^3)}{\chi(-q)}.$$

Therefore, by (5.7.6), (5.7.7), and (5.7.8), we find that

(5.7.9)
$$\chi(q)\chi(-q^3) = \frac{f(q,q^5)}{f(-q^{12})} = \frac{f(q^8,q^{16})}{f(-q^{12})} + q\frac{f(q^4,q^{20})}{f(-q^{12})}$$
$$= \frac{\varphi(-q^{24})}{\chi(-q^8)f(-q^{12})} + q\frac{\chi(q^4)\chi(-q^{12})f(-q^{48})}{f(-q^{12})}.$$

Next, by several applications of (2.14), we deduce from (5.7.9) that

(5.7.10) $$\chi(q)\chi(-q^3) = \frac{\chi(q^{12})}{\chi(-q^8)} + q\frac{\chi(q^4)}{\chi(-q^{24})}.$$

Therefore, by (5.7.5), (5.7.10), and (2.16),

(5.7.11)
$$G(q^{16})G(q^{24}) + q^8 H(q^{16})H(q^{24}) + q\left(H(-q^4)G(q^{24}) + q^4 G(-q^4)H(q^{24})\right)$$
$$= \frac{\chi(q^{12})\chi(-q^{12})}{\chi(-q^8)} + q\frac{\chi(q^4)\chi(-q^{12})}{\chi(-q^{24})} = \frac{\chi(-q^{24})}{\chi(-q^8)} + q\frac{\chi(q^4)}{\chi(q^{12})}.$$

Equating the even parts on both sides of the equation (5.7.11), we obtain Entry 3.7 with q replaced by q^8. Similarly, equating the odd parts gives Entry 3.8 with q replaced by $-q^4$. □

5.8. Proof of Entry 3.8

We have just given a proof of Entry 3.8 along with one of our proofs of Entry 3.7. We provide a second proof here.

If we use (2.11), we can put Entry 3.8 in the form

(5.8.1)
$$f(-q^{12}, -q^{18})f(-q, -q^4) - qf(-q^2, -q^3)f(-q^6, -q^{24}) = \frac{\chi(-q)}{\chi(-q^3)}f(-q)f(-q^6).$$

Using (2.7)–(2.10), we can rewrite (5.8.1) as

(5.8.2) $\quad f(-q^{12}, -q^{18})f(-q, -q^4) - qf(-q^2, -q^3)f(-q^6, -q^{24}) = \psi(q^3)\varphi(-q).$

In the representation,

(5.8.3) $\quad 2\psi(q^3)\varphi(-q) = f(1, q^3)f(-q, -q) = \sum_{m,n=-\infty}^{\infty} (-1)^n q^{(3m^2+3m+2n^2)/2},$

we make the change of indices

(5.8.4) $\qquad 3m - 2n = 5M + a \qquad$ and $\qquad m + n = 5N + b,$

where a and b have values selected from the integers $0, \pm 1, \pm 2$. Since

(5.8.5) $\quad m = M + 2N + (a + 2b)/5 \qquad$ and $\qquad n = -M + 3N + (3b - a)/5,$

we see that values of a and b are associated as in the following table:

a	0	± 1	± 2
b	0	± 2	∓ 1

Thus, there is a one-one correspondence between all pairs of integers (m,n) and all sets of integers (M,N,a) as given above. We therefore deduce from (5.8.3), (2.4), and (2.5) that

$$2\psi(q^3)\varphi(-q) = \sum_{m,n=-\infty}^{\infty} (-1)^n q^{(3m^2+3m+2n^2)/2}$$
$$= -qf(-q^2,-q^3)f(-q^6,-q^{24}) - qf(-q^2,-q^3)f(-q^6,-q^{24})$$
$$+ f(-q,-q^4)f(-q^{12},-q^{18}) - q^4 f(-1,-q^5)f(-1,-q^{30})$$
$$- qf(-q^6,-q^{-1})f(-q^{12},-q^{18})$$
$$= -2qf(-q^2,-q^3)f(-q^6,-q^{24}) + f(-q,-q^4)f(-q^{12},-q^{18})$$
$$+ f(-q,-q^4)f(-q^{12},-q^{18})$$
$$= 2f(-q,-q^4)f(-q^{12},-q^{18}) - 2qf(-q^2,-q^3)f(-q^6,-q^{24}).$$

By (5.8.2), we see that the proof is complete.

5.9. Proof of Entry 3.9

By using (2.11) and (2.14), we find that Entry 3.9 is equivalent to the identity

$$f(-q^{14},-q^{21})f(-q^2,-q^8) - qf(-q^4,-q^6)f(-q^7,-q^{28}) = f(-q^2)f(-q^7)\frac{\chi(-q)}{\chi(-q^7)}$$
(5.9.1)
$$= f(-q)f(-q^{14}).$$

We invoke (4.13) with the two sets of parameters $\alpha_1 = 2, \beta_1 = 7, m_1 = 3, p_1 = 5, \lambda_1 = 5$ and $\alpha_2 = 1, \beta_2 = 14, m_2 = 1, p_2 = 3, \lambda_2 = 5$. The conditions in (4.9) are easily seen to be met. By using (4.14) and (4.15), we find that

$$f(-q^{28},-q^{42})f(-q^4,-q^{16}) - q^2 f(-q^8,-q^{12})f(-q^{14},-q^{56}) = f(-q^2)f(-q^{28}).$$

Replacing q^2 by q in the last equality, we deduce (5.9.1) to complete the proof.

5.10. Proof of Entry 3.10

By using (2.11) and (2.14), we find that Entry 3.10 is equivalent to the identity

$$f(-q^2,-q^3)f(-q^{28},-q^{42}) + q^3 f(-q,-q^4)f(-q^{14},-q^{56}) = f(-q)f(-q^{14})\frac{\chi(-q^7)}{\chi(-q)}$$
(5.10.1)
$$= f(-q^2)f(-q^7).$$

We now apply (4.13) with $\alpha_1 = 1, \beta_1 = 14, m_1 = 1, p_1 = 5, \lambda_1 = 3$ and $\alpha_2 = 2, \beta_2 = 7, m_2 = 1, p_2 = 3, \lambda_2 = 3$. We easily find that these sets of parameters satisfy the conditions in (4.9). Employing (4.13) and (4.15), we find that
(5.10.2)
$$f(-q^4,-q^6)f(-q^{56},-q^{84}) + q^6 f(-q^2,-q^8)f(-q^{28},-q^{112}) = f(-q^4)f(-q^{14}),$$

which is (5.10.1), but with q replaced by q^2, and so the proof is complete.

5.11. Proofs of Entry 3.11

First Proof of Entry 3.11. By employing (2.11), we see that Entry 3.11 is equivalent to the identity

(5.11.1) $f(-q^{16}, -q^{24})f(-q^3, -q^{12}) - qf(-q^6, -q^9)f(-q^8, -q^{32})$
$$= f(-q^3)f(-q^8)\frac{\chi(-q)\chi(-q^4)}{\chi(-q^3)\chi(-q^{12})}.$$

We apply (4.13) with the two sets of parameters $\alpha_1 = 3, \beta_1 = 8, m_1 = 3, p_1 = 5, \lambda_1 = 7$ and $\alpha_2 = 2, \beta_2 = 12, m_2 = 1, p_2 = 2, \lambda_2 = 7$. The conditions (4.9) are easily seen to be satisfied. Using (4.14) and (4.16), we find that
(5.11.2)
$$f(-q^{32}, -q^{48})f(-q^6, -q^{24}) - q^2 f(-q^{12}, -q^{18})f(-q^{16}, -q^{64}) = \psi(-q^2)\psi(-q^{12}).$$

After replacing q^2 by q in (5.11.2) and comparing the result with (5.11.1), we find that it suffices to show that

(5.11.3) $$\frac{\psi(-q)\psi(-q^6)}{f(-q^3)f(-q^8)} = \frac{\chi(-q)\chi(-q^4)}{\chi(-q^3)\chi(-q^{12})}.$$

By using the product representations for $\psi(-q^a)$ and $f(-q^a)$ in (2.8) and (2.9), respectively, we find that

$$\frac{\psi(-q)\psi(-q^6)}{f(-q^3)f(-q^8)} = \frac{(q^2;q^2)_\infty (q^{12},q^{12})_\infty}{(-q;q^2)_\infty(-q^6,q^{12})_\infty(q^3;q^3)_\infty(q^8;q^8)_\infty}$$
$$= \frac{(q^2;q^2)_\infty(q;q^2)_\infty(q^{12},q^{12})_\infty(q^6;q^{12})_\infty}{(q^2;q^4)_\infty(q^{12},q^{24})_\infty(q^3;q^3)_\infty(q^8;q^8)_\infty}$$
$$= \frac{\chi(-q)\chi(-q^4)}{\chi(-q^{12})}\frac{(q^6;q^6)_\infty}{(q^3;q^3)_\infty}$$
$$= \frac{\chi(-q)\chi(-q^4)}{\chi(-q^3)\chi(-q^{12})}.$$

Thus, the proof of (5.11.3) is complete, and so is that of Entry 3.11 also complete. □

Second Proof of Entry 3.11. Using (4.23) and (4.24) in (3.9), we find that

$$\frac{\chi(-q)}{\chi(-q^3)} = G(q^6)H(q) - qG(q)H(q^6)$$
$$= \frac{f(-q^8)}{f(-q^2)}\left\{G(q^6)\left(q^3 H(q^{16}) + G(-q^4)\right) - qH(q^6)\left(G(q^{16}) + qH(-q^4)\right)\right\}$$
$$= \frac{f(-q^8)}{f(-q^2)}\Big\{G(-q^4)G(q^6) - q^2 H(-q^4)H(q^6)$$
(5.11.4) $\qquad - q\left(H(q^6)G(q^{16}) - q^2 G(q^6)H(q^{16})\right)\Big\}.$

By (2.15) and (2.17), and by (5.7.10) with q replaced by $-q$, we deduce from (5.11.4) that

(5.11.5)
$$G(-q^4)G(q^6) - q^2 H(-q^4)H(q^6) - q\left(H(q^6)G(q^{16}) - q^2 G(q^6)H(q^{16})\right)$$
$$= \frac{\chi(-q)f(-q^2)}{\chi(-q^3)f(-q^8)} = \frac{f(-q^2)}{f(-q^8)\chi(-q^6)}\chi(-q)\chi(q^3)$$
$$= \frac{f(-q^2)}{f(-q^8)\chi(-q^6)}\left\{\frac{\chi(q^{12})}{\chi(-q^8)} - q\frac{\chi(q^4)}{\chi(-q^{24})}\right\} = \frac{\chi(-q^2)\chi(q^{12})}{\chi(q^4)\chi(-q^6)} - q\frac{\chi(-q^2)\chi(-q^8)}{\chi(-q^6)\chi(-q^{24})}.$$

Equating the even and odd parts on both sides of the equation (5.11.5), we obtain Entries 3.23 and 3.11 with q replaced by $-q^2$ and q^2, respectively. □

5.12. Proofs of Entry 3.12

The first proof that we give is due to Bressoud [14].

First Proof of Entry 3.12. Using (2.11), we readily find that Entry 3.12 is equivalent to the identity

(5.12.1) $\quad f(-q^2, -q^3)f(-q^{48}, -q^{72}) + q^5 f(-q, -q^4)f(-q^{24}, -q^{96})$
$$= f(-q)f(-q^{24})\frac{\chi(-q^3)\chi(-q^{12})}{\chi(-q)\chi(-q^4)}.$$

We apply (4.13) with the two sets of parameters $\alpha_1 = 1, \beta_1 = 24, m_1 = 1, p_1 = 5, \lambda_1 = 5$ and $\alpha_2 = 4, \beta_2 = 6, m_2 = 1, p_2 = 2, \lambda_2 = 5$. We find that the conditions in (4.9) are satisfied. Hence, using (4.15) and (4.18), we deduce the identity

(5.12.2)
$$f(-q^4, -q^6)f(-q^{96}, -q^{144}) + q^{10}f(-q^2, -q^8)f(-q^{48}, -q^{192}) = \psi(-q^4)\psi(-q^6).$$

Replacing q^2 by q in (5.12.2) and comparing it with (5.12.1), we find that it suffices to prove that

(5.12.3) $$\frac{\psi(-q^2)\psi(-q^3)}{f(-q)f(-q^{24})} = \frac{\chi(-q^3)\chi(-q^{12})}{\chi(-q)\chi(-q^4)}.$$

Using the product representations of $\psi(-q^a)$ and $f(-q^a)$ from (2.8) and (2.9), respectively, we find that

$$\frac{\psi(-q^2)\psi(-q^3)}{f(-q)f(-q^{24})} = \frac{(q^4;q^4)_\infty (q^6;q^6)_\infty}{(-q^2;q^4)_\infty(-q^3;q^6)_\infty(q;q)_\infty(q^{24};q^{24})_\infty}$$
$$= \frac{(q^4;q^4)_\infty(q^6;q^6)_\infty(q^2;q^4)_\infty(q^3;q^6)_\infty}{(q^4;q^8)_\infty(q^6;q^{12})_\infty(q;q)_\infty(q^{24};q^{24})_\infty}$$
$$= \frac{(q^2;q^2)_\infty(q^6;q^6)_\infty\chi(-q^3)}{\chi(-q^4)(q;q)_\infty(q^6;q^{12})_\infty(q^{24};q^{24})_\infty}$$
$$= \frac{(q^6;q^6)_\infty\chi(-q^3)}{\chi(-q)\chi(-q^4)(q^6;q^{12})_\infty(q^{24};q^{24})_\infty}$$
$$= \frac{\chi(-q^3)\chi(-q^{12})}{\chi(-q)\chi(-q^4)},$$

which establishes (5.12.3), and so the proof is complete. □

Second Proof of Entry 3.12. Using (4.23) and (4.24) in (3.8) with q replaced by q^3, we arrive at

(5.12.4)
$$\frac{\chi(-q^3)}{\chi(-q)} = G(q^2)G(q^3) + qH(q^2)H(q^3)$$
$$= \frac{f(-q^{24})}{f(-q^6)} \left\{ G(q^2)\left(G(q^{48}) + q^3 H(-q^{12})\right) + qH(q^2)\left(q^9 H(q^{48}) + G(-q^{12})\right) \right\}$$
$$= \frac{f(-q^{24})}{f(-q^6)} \left\{ G(q^2)G(q^{48}) + q^{10} H(q^2)H(q^{48}) \right.$$
$$\left. + q\left(H(q^2)G(-q^{12}) + q^2 G(q^2)H(-q^{12})\right) \right\}.$$

That is,

(5.12.5) $G(q^2)G(q^{48}) + q^{10} H(q^2)H(q^{48}) + q\left(H(q^2)G(-q^{12}) + q^2 G(q^2)H(-q^{12})\right)$
$$= \frac{f(-q^6)\chi(-q^3)}{f(-q^{24})\chi(-q)}.$$

Therefore, by (5.7.10), (2.15), and (2.14),
(5.12.6)
$$\frac{f(-q^6)\chi(-q^3)}{f(-q^{24})\chi(-q)} = \frac{f(-q^6)\chi(q)\chi(-q^3)}{f(-q^{24})\chi(-q^2)} = \frac{\chi(-q^6)\chi(-q^{24})}{\chi(-q^2)\chi(-q^8)} + q\frac{\chi(q^4)\chi(-q^6)}{\chi(-q^2)\chi(q^{12})}.$$

Returning to (5.12.5), we use (5.12.6) to equate the odd parts on both sides of the equation, and, upon replacing q^2 by $-q$, we find that
$$H(-q)G(-q^6) - qG(-q)H(-q^6) = \frac{\chi(q^2)\chi(q^3)}{\chi(q)\chi(q^6)},$$
which is Entry 3.24. Similarly, equating the even parts in (5.12.5), employing (5.12.6), and replacing q^2 by q, we deduce that
$$G(q)G(q^{24}) + q^5 H(q)H(q^{24}) = \frac{\chi(-q^3)\chi(-q^{12})}{\chi(-q)\chi(-q^4)},$$
which is Entry 3.12. □

5.13. Proofs of Entries 3.13 and 3.14

Throughout this section, we shall use several times without comment the elementary identity (2.5). To prove Entries 3.13 and 3.14, we also need the following lemma.

LEMMA 5.4. *For* $|q| < 1$, $x \neq 0$, *and* $y \neq 0$,
$$f(-x, -x^{-1}q)f(-y, -y^{-1}q)$$
(5.13.1) $= f(xy, (xy)^{-1}q^2)f(x^{-1}yq, xy^{-1}q) - xf(xyq, (xy)^{-1}q)f(x^{-1}y, xy^{-1}q^2).$

In this form, Lemma 5.4 is given as Theorem 1.1 in [**18**, p. 649]. However, it is easily seen that Lemma 5.4 can be obtained by adding Entries 29(i) and (ii) in Chapter 16 of Ramanujan's second notebook [**5**, p. 45].

LEMMA 5.5.
(5.13.2) $\quad f^2(-q^{27}, -q^{45}) + q^9 f^2(-q^9, -q^{63}) = f(-q^3, q^6)\psi(-q^9)\chi(q^3).$

Proof. Replacing q, x, and y by q^{36}, $-q^9$, and q^{18}, respectively, in Lemma 5.4, we find that
(5.13.3)
$$f(-q^{27},-q^{45})^2 + q^9 f(-q^9,-q^{63})^2 = f(q^9,q^{27})f(-q^{18},-q^{18}) = \psi(q^9)\varphi(-q^{18}).$$

Using (2.13)–(2.15), or using (2.7)–(2.9), we can easily conclude that
(5.13.4)
$$\varphi(-q^2)\psi(q) = \varphi(q)\psi(-q).$$

Therefore, by (5.7.8) with q replaced by $-q^3$, and by (5.13.4) with q replaced by q^9, we find that
(5.13.5)
$$f(-q^3,q^6)\psi(-q^9)\chi(q^3) = \psi(-q^9)\varphi(q^9) = \psi(q^9)\varphi(-q^{18}).$$

Thus, we have proved Lemma 5.5. □

LEMMA 5.6.
$$f(-q^{21},-q^{51})f(-q^{27},-q^{45}) + q^6 f(-q^3,-q^{69})f(-q^{27},-q^{45})$$
$$+ q^3 f(-q^{33},-q^{39})f(-q^9,-q^{63}) - q^6 f(-q^{15},-q^{57})f(-q^9,-q^{63})$$
(5.13.6) $\qquad = \psi^2(-q^9)\chi(q^3).$

Proof. Replacing q, x, and y by q^{36}, q^6, and $-q^{15}$, respectively, in Lemma 5.4, we find that
$$f(-q^{21},-q^{51})f(-q^{27},-q^{45}) - q^6 f(-q^{15},-q^{57})f(-q^9,-q^{63})$$
(5.13.7) $\qquad = f(-q^6,-q^{30})f(q^{15},q^{21}),$

and replacing q, x, and y by q^{36}, $-q^3$, and q^6, respectively, in Lemma 5.4, we find that
$$f(-q^9,-q^{63})f(-q^{33},-q^{39}) + q^3 f(-q^{27},-q^{45})f(-q^3,-q^{69})$$
(5.13.8) $\qquad = f(q^3,q^{33})f(-q^6,-q^{30}).$

By (2.19) with $a = q^3$, $b = q^6$ and $n = 2$, we deduce that
(5.13.9)
$$f(q^3,q^6) = f(q^{15},q^{21}) + q^3 f(q^3,q^{33}).$$

Thus, by (5.13.7)–(5.13.9), we find that
$$f(-q^{21},-q^{51})f(-q^{27},-q^{45}) + q^6 f(-q^3,-q^{69})f(-q^{27},-q^{45})$$
$$+ q^3 f(-q^{33},-q^{39})f(-q^9,-q^{63}) - q^6 f(-q^{15},-q^{57})f(-q^9,-q^{63})$$
(5.13.10) $\qquad = f(q^3,q^6)f(-q^6,-q^{30}).$

One can easily verify that
(5.13.11)
$$\psi^2(-q) = \psi(q^2)\varphi(-q).$$

By (5.7.8), (5.7.6), and (5.13.11) with q replaced by q^3, q^6, and q^9, respectively, and by (2.15), we conclude that
$$f(q^3,q^6)f(-q^6,-q^{30}) = \frac{\varphi(-q^9)}{\chi(-q^3)}\chi(-q^6)\psi(q^{18})$$
(5.13.12) $\qquad = \varphi(-q^9)\psi(q^{18})\chi(q^3) = \psi^2(-q^9)\chi(q^3).$

Thus, we have proved Lemma 5.6. □

THEOREM 5.1. *For $|q|<1$,*

$$f(-q,-q^7)f(-q^{27},-q^{45}) - q^4 f(-q^3,-q^5)f(-q^9,-q^{63})$$
(5.13.13)
$$= (q^4;q^4)_\infty (q^9;q^9)_\infty \frac{\chi(-q)\chi(-q^6)}{\chi(-q^3)\chi(-q^{18})}.$$

Proof. Replacing n, a, and b by 3, $-q$, and $-q^7$, respectively, in (2.19), we find that
(5.13.14) $$f(-q,-q^7) = f(-q^{27},-q^{45}) - qf(-q^{21},-q^{51}) - q^7 f(-q^3,-q^{69});$$
replacing n, a, and b by 3, $-q^3$, and $-q^5$, respectively, in (2.19), we find that
(5.13.15) $$f(-q^3,-q^5) = f(-q^{33},-q^{39}) - q^3 f(-q^{15},-q^{57}) - q^5 f(-q^9,-q^{63});$$
and replacing n, a, and b by 3, $-q$, and $-q^3$, respectively, in (2.19), we find that (see also [**5**, p. 49, Cor.])

$$\psi(-q) = f(-q,-q^3) = f(-q^{15},-q^{21}) - qf(-q^9,-q^{27}) - q^3 f(-q^3,-q^{33})$$
(5.13.16) $$= f(-q^3, q^6) - q\psi(-q^9),$$

where in the last step, we used (5.13.9) with q replaced by $-q$. Then, by (5.13.14) and (5.13.15), the left-hand side of (5.13.13) equals

$$f^2(-q^{27},-q^{45}) - qf(-q^{21},-q^{51})f(-q^{27},-q^{45})$$
$$- q^7 f(-q^3,-q^{69})f(-q^{27},-q^{45}) - q^4 f(-q^{33},-q^{39})f(-q^9,-q^{63})$$
(5.13.17) $$+ q^7 f(-q^{15},-q^{57})f(-q^9,-q^{63}) + q^9 f^2(-q^9,-q^{63}).$$

Therefore, by (5.13.17), Lemma 5.5, Lemma 5.6, (5.13.16), (2.16) and (2.17) the left-hand side of (5.13.13) equals

$$f(-q^3, q^6)\psi(-q^9)\chi(q^3) - q\psi^2(-q^9)\chi(q^3)$$
$$= \psi(-q^9)\chi(q^3)\left\{f(-q^3,q^6) - q\psi(-q^9)\right\}$$
(5.13.18) $$= \psi(-q)\psi(-q^9)\chi(q^3) = f(-q^4)\chi(-q)\frac{f(-q^9)}{\chi(-q^{18})}\frac{\chi(-q^6)}{\chi(-q^3)}.$$

We have thus completed the proof of Theorem 5.1. □

THEOREM 5.2. *For $|q|<1$,*

$$f(-q^4,-q^{16})f(-q^{18},-q^{27}) - qf(-q^8,-q^{12})f(-q^9,-q^{36})$$
$$= f(-q,-q^4)f(-q^{72},-q^{108}) - q^7 f(-q^2,-q^3)f(-q^{36},-q^{144})$$
(5.13.19) $$= f(-q,-q^7)f(-q^{27},-q^{45}) - q^4 f(-q^3,-q^5)f(-q^9,-q^{63}).$$

Proof. We apply the ideas of Rogers with the three sets of parameters $\alpha_1 = 4$, $\beta_1 = 9$, $m_1 = 3$, $p_1 = 5$, $\lambda_1 = 9$; $\alpha_2 = 1$, $\beta_2 = 36$, $m_2 = 3$, $p_2 = 5$, $\lambda_2 = 9$; and $\alpha_3 = 2$, $\beta_3 = 18$, $m_3 = 3$, $p_3 = 4$, $\lambda_3 = 9$. The requisite conditions (4.9) are satisfied. Therefore, we find that

$$q^{9/4}f(-q^8,-q^{32})f(-q^{36},-q^{54}) - q^{17/4}f(-q^{16},-q^{24})f(-q^{18},-q^{72})$$
$$= q^{9/4}f(-q^2,-q^8)f(-q^{144},-q^{216}) - q^{65/4}f(-q^4,-q^6)f(-q^{72},-q^{288})$$
(5.13.20) $$= q^{9/4}f(-q^2,-q^{14})f(-q^{54},-q^{90}) - q^{41/4}f(-q^6,-q^{10})f(-q^{18},-q^{126}).$$

Dividing each term of (5.13.20) by $q^{9/4}$, and replacing q^2 by q, we are able to derive (5.13.19) from (5.13.20). □

We are now going to prove Entries 3.13 and 3.14.

Proof of Entry 3.13. By Theorems 5.1 and 5.2, we find that
$$f(-q^4,-q^{16})f(-q^{18},-q^{27}) - qf(-q^8,-q^{12})f(-q^9,-q^{36})$$
(5.13.21)
$$= (q^4;q^4)_\infty (q^9;q^9)_\infty \frac{\chi(-q)\chi(-q^6)}{\chi(-q^3)\chi(-q^{18})}.$$

Dividing both sides of (5.13.21) by $(q^4;q^4)_\infty (q^9;q^9)_\infty$ and using the definitions of $G(q)$ and $H(q)$, we derive Entry 3.13. □

Proof of Entry 3.14. By Theorems 5.1 and 5.2, we find that
$$f(-q,-q^4)f(-q^{72},-q^{108}) - q^7 f(-q^2,-q^3)f(-q^{36},-q^{144})$$
(5.13.22)
$$= (q^4;q^4)_\infty (q^9;q^9)_\infty \frac{\chi(-q)\chi(-q^6)}{\chi(-q^3)\chi(-q^{18})}.$$

Dividing both sides of (5.13.22) by $(q;q)_\infty (q^{36};q^{36})_\infty$ and using the definitions of $G(q)$ and $H(q)$, we find that the left-hand side of (5.13.22) equals $G(q^{36})H(q) - q^7 G(q)H(q^{36})$, and the right-hand side of (5.13.22) equals
$$\frac{(q^4;q^4)_\infty (q^9;q^9)_\infty}{(q;q)_\infty (q^{36};q^{36})_\infty} \cdot \frac{\chi(-q)\chi(-q^6)}{\chi(-q^3)\chi(-q^{18})} = \frac{\chi(-q^6)\chi(-q^9)}{\chi(-q^2)\chi(-q^3)},$$
which completes the proof of Entry 3.14. □

We offer now a second, completely different proof of Entry 3.13.

Second Proof of Entry 3.13. By (2.11), (2.8), (2.6), and some elementary product manipulations, Entry 3.13 is easily seen to be equivalent to
(5.13.23)
$$f(-q^4,-q^{16})f(-q^{18},-q^{27}) - qf(-q^8,-q^{12})f(-q^9,-q^{36}) = \psi(-q)f(q^3,q^{15}).$$
We prove (5.13.23).

Employing Theorem 4.1 with the set of parameters $a = q^3$, $b = q^{15}$, $c = q$, $d = q^2$, $\alpha = 3$, $\beta = 1$, $m = 5$, $\epsilon_1 = 0$, and $\epsilon_2 = 1$, we find that
$$f(q^3,q^{15})f(-q,-q^2) = f(-q^{18},-q^{27})f(q^8,q^{22}) - qf(-q^9,-q^{36})f(q^{14},q^{16})$$
$$+ q^3 f(-q^9,-q^{36})f(q^4,q^{26}) - q^2 f(-q^{18},-q^{27})f(q^2,q^{28}).$$

Upon the rearrangement of terms and the use of (4.28) and (4.29) with q replaced by q^2, we deduce that
$$f(q^3,q^{15})f(-q) = f(-q^{18},-q^{27})\{f(q^8,q^{22}) - q^2 f(q^2,q^{28})\}$$
$$- qf(-q^9,-q^{36})\{f(q^{14},q^{16}) - q^2 f(q^4,q^{26})\}$$
$$= f(-q^{18},-q^{27})H(q^4)f(-q^2) - qf(-q^9,-q^{36})G(q^4)f(-q^2)$$
(5.13.24)
$$= \frac{f(-q^2)}{f(-q^4)}\{f(-q^4,-q^{16})f(-q^{18},-q^{27}) - qf(-q^8,-q^{12})f(-q^9,-q^{36})\}.$$

But by (2.17),
$$\frac{f(-q)f(-q^4)}{f(-q^2)} = \psi(-q).$$

Using the last equality in (5.13.24), we complete the proof (5.13.23) and also that of Entry 3.14. □

In Section 5.6, we promised that in the current section we would give a proof that simultaneously yields Entries 3.6 and 3.13. We show that Entry 3.11 implies both Entries 3.6 and 3.13.

Another Proof of Entries 3.6 and 3.13. In Entry 3.11, we employ (4.31) and (4.30) with q replaced by q^3 to find that

$$\left\{-q^3\frac{\chi(q^{18})}{\chi(-q^{12})}H(-q^{18}) + \frac{\chi(q^6)}{\chi(-q^{36})}G(q^{72})\right\}G(q^8)$$
$$-q\left\{\frac{\chi(q^{18})}{\chi(-q^{12})}G(-q^{18}) - q^{15}\frac{\chi(q^6)}{\chi(-q^{36})}H(q^{72})\right\}H(q^8)$$
(5.13.25) $$= \chi(q^9)\chi(-q^3)\frac{\chi(-q)\chi(-q^4)}{\chi(-q^3)\chi(-q^{12})}.$$

Upon collecting terms, we deduce from (5.13.25) that

(5.13.26)
$$\frac{\chi(q^6)}{\chi(-q^{36})}\left\{G(q^8)G(q^{72}) + q^{16}H(q^8)H(q^{72})\right\}$$
$$-q\frac{\chi(q^{18})}{\chi(-q^{12})}\left\{G(-q^{18})H(q^8) + q^2H(-q^{18})G(q^8)\right\} = \frac{\chi(-q^4)}{\chi(-q^{12})}\chi(-q)\chi(q^9).$$

To equate even and odd parts on both sides of (5.13.26), we need the 2-dissection of $\chi(-q)\chi(q^9)$ which we obtain from Theorem 4.1. To that end, we employ Theorem 4.1 with the set of parameters $a=1$, $b=q^9$, $c=q$, $d=q^2$, $\epsilon_1=0$, $\epsilon_2=1$, $\alpha=\beta=1$, and $m=4$ to find that

$$f(1,q^9)f(-q,-q^2) = f(-q^2,-q^{10})f(-q^{12},-q^{24}) + f(-q,-q^{11})f(-q^{15},-q^{21})$$
(5.13.27) $$\qquad -qf(-q^4,-q^8)f(-q^6,-q^{30}) - q^2f(-q^5,-q^7)f(-q^3,-q^{33}).$$

We employ Theorem 4.1 again with the same set of parameters, except this time we take $\epsilon_1=1$, and $\epsilon_2=0$, to find that

$$f(-1,-q^9)f(q,q^2) = f(-q^2,-q^{10})f(-q^{12},-q^{24}) - f(-q,-q^{11})f(-q^{15},-q^{21})$$
(5.13.28) $$\qquad -qf(-q^4,-q^8)f(-q^6,-q^{30}) + q^2f(-q^5,-q^7)f(-q^3,-q^{33}).$$

By (2.4), the product on the left side of (5.13.28) equals 0. Recalling the definitions (2.8) and (2.9), and employing (2.3), (5.13.27), and (5.13.28), we conclude that

$$\psi(q^9)f(-q) = \frac{1}{2}f(1,q^9)f(-q,-q^2) = \frac{1}{2}\left\{f(1,q^9)f(-q,-q^2) + f(-1,-q^9)f(q,q^2)\right\}$$
$$= f(-q^2,-q^{10})f(-q^{12},-q^{24}) - qf(-q^4,-q^8)f(-q^6,-q^{30})$$
(5.13.29) $$= f(-q^2,-q^{10})f(-q^{12}) - qf(-q^4)f(-q^6,-q^{30}).$$

Next, we use (2.14), (2.17), and (5.7.6) twice with q replaced by q^2 and q^6, respectively, to find from (5.13.29) that

$$\chi(q^9)f(-q^{36})\chi(-q)f(-q^2)$$
$$= f(-q^{12})\chi(-q^2)\chi(q^6)f(-q^{24}) - qf(-q^4)\chi(-q^6)\chi(q^{18})f(-q^{72}),$$

which after several uses of (2.14) simplifies to

(5.13.30) $$\chi(-q)\chi(q^9) = \frac{f(-q^{12})\psi(q^6)}{f(-q^4)f(-q^{36})} - q\frac{\chi(-q^6)}{\chi(-q^2)\chi(-q^{18})}.$$

Returning to (5.13.26), we substitute the value of $\chi(-q)\chi(q^9)$ from (5.13.30) and equate the odd parts on both sides of the resulting equation. Hence, using (2.15), we conclude that

$$G(-q^{18})H(q^8) + q^2H(-q^{18})G(q^8) = \frac{\chi(-q^{12})}{\chi(q^{18})}\frac{\chi(-q^4)}{\chi(-q^{12})}\frac{\chi(-q^6)}{\chi(-q^2)\chi(-q^{18})}$$
$$= \frac{\chi(q^2)\chi(-q^6)}{\chi(-q^{36})},$$

which is Entry 3.13 with q replaced by $-q^2$. Similarly, equating the even parts in (5.13.26) with the use of (5.13.30), using (2.14) and (2.17), and replacing q^8 by q, we deduce Entry 3.6. □

5.14. Proof of Entry 3.15

Let

(5.14.1) $$M(q) := G(q^3)G(q^7) + q^2H(q^3)H(q^7)$$

and

(5.14.2) $$N(q) := G(q^{21})H(q) - q^4G(q)H(q^{21}).$$

Consider the system of equations

$$N(q^2) = H(q^2)G(q^{42}) - q^8G(q^2)H(q^{42}),$$

(5.14.3) $$\frac{\chi(-q^7)}{\chi(-q^{21})} =: R(q) = H(q^7)G(q^{42}) - q^7G(q^7)H(q^{42}),$$

(5.14.4) $$\frac{\chi(-q^{21})}{\chi(-q^3)} =: S(q) = G(q^3)G(q^{42}) + q^9H(q^3)H(q^{42}),$$

arising from (5.14.2), Entry 3.8 with q replaced by q^7, and Entry 3.10 with q replaced by q^3, respectively. It follows that

$$\begin{vmatrix} H(q^2) & -q^8G(q^2) & N(q^2) \\ H(q^7) & -q^7G(q^7) & R(q) \\ G(q^3) & q^9H(q^3) & S(q) \end{vmatrix} = 0,$$

or, using (5.14.1), Entry 3.7, Entry 3.9, (5.14.3), and (5.14.4), we find that

$$0 = N(q^2)\left(q^9H(q^3)H(q^7) + q^7G(q^3)G(q^7)\right)$$
$$- R(q)\left(q^9H(q^2)H(q^3) + q^8G(q^2)G(q^3)\right)$$
$$+ S(q)\left(-q^7G(q^7)H(q^2) + q^8G(q^2)H(q^7)\right)$$
$$= q^7N(q^2)M(q) - q^8\frac{\chi(-q^7)}{\chi(-q^{21})}\frac{\chi(-q^3)}{\chi(-q)} - q^7\frac{\chi(-q^{21})}{\chi(-q^3)}\frac{\chi(-q)}{\chi(-q^7)}.$$

Solving the equation above for $N(q^2)M(q)$, we find that, if

(5.14.5) $$T(q) := \frac{\chi(-q^3)\chi(-q^7)}{\chi(-q)\chi(-q^{21})},$$

then

(5.14.6) $$N(q^2)M(q) = qT(q) + \frac{1}{T(q)}.$$

Next, we derive a similar formula for $M(q^2)N(q)$. Using (5.14.1), Entry 3.10, and Entry 3.7 with q replaced by q^7, we find that

$$M(q^2) = G(q^6)G(q^{14}) + q^4 H(q^6)H(q^{14}),$$

(5.14.7) $\quad \dfrac{\chi(-q^7)}{\chi(-q)} =: R_1(q) = G(q)G(q^{14}) + q^3 H(q)H(q^{14}),$

(5.14.8) $\quad \dfrac{\chi(-q^{21})}{\chi(-q^7)} =: S_1(q) = G(q^{21})G(q^{14}) + q^7 H(q^{21})H(q^{14}),$

which implies that

$$\begin{vmatrix} G(q^6) & q^4 H(q^6) & M(q^2) \\ G(q) & q^3 H(q) & R_1(q) \\ G(q^{21}) & q^7 H(q^{21}) & S_1(q) \end{vmatrix} = 0.$$

Hence, by (5.14.2), Entry 3.9 with q replaced by q^3, Entry 3.8, (5.14.7), and (5.14.8),

$$\begin{aligned} 0 &= M(q^2)\left(q^7 G(q)H(q^{21}) - q^3 H(q)G(q^{21})\right) \\ &\quad - R_1(q)\left(q^7 G(q^6)H(q^{21}) - q^4 H(q^6)G(q^{21})\right) \\ &\quad + S_1(q)\left(q^3 G(q^6)H(q) - q^4 H(q^6)G(q)\right) \\ &= -q^3 M(q^2)N(q) + q^4 \frac{\chi(-q^7)}{\chi(-q)} \frac{\chi(-q^3)}{\chi(-q^{21})} + q^3 \frac{\chi(-q^{21})}{\chi(-q^7)} \frac{\chi(-q)}{\chi(-q^3)}. \end{aligned}$$

Hence, solving the equation above for $M(q^2)N(q)$, we find that

(5.14.9) $\quad M(q^2)N(q) = qT(q) + \dfrac{1}{T(q)},$

where $T(q)$ is defined by (5.14.5). Comparing (5.14.6) with (5.14.9), we find that

(5.14.10) $\quad N(q^2)M(q) = M(q^2)N(q).$

Equation (5.14.10) easily implies that $M(q) = N(q)$, which is what we wanted to prove, i.e., (3.16). To see this, let

$$M(q) := \sum_{n=0}^{\infty} a_n q^n \quad \text{and} \quad N(q) := \sum_{n=0}^{\infty} b_n q^n.$$

From the definitions (5.14.1) and (5.14.2), we see that $a_0 = b_0$. Then by an easy inductive argument, we find that $a_n = b_n$, for every positive integer n. Hence, $M(q) = N(q)$, as we wanted to demonstrate.

As an immediate consequence of our main identity $M(q) = N(q)$ and (5.14.9), we derive the following curious corollary.

COROLLARY 5.1. *If $T(q)$ is defined by (5.14.5), then*

$$M(q)M(q^2) = N(q)N(q^2) = qT(q) + \dfrac{1}{T(q)}.$$

Next, we prove the second part of Entry 3.15, i.e., (3.17). Let $J(q)$ denote the right-hand side of (3.17), so that

(5.14.11) $\quad J(q^2) = \dfrac{1}{2q}\left\{\chi(q)\chi(-q^3)\chi(q^7)\chi(-q^{21}) - \chi(-q)\chi(q^3)\chi(-q^7)\chi(q^{21})\right\}.$

Recall that $M(q)$ and $N(q)$ are defined by (5.14.1) and (5.14.2), respectively. Using the previously established fact, $M(q) = N(q)$, we see that it suffices to show that $M(q^2)N(q^2) = J^2(q^2)$.

Using (4.39) and (4.40) in (3.10) with q replaced by q^7, we obtain

$$\{a(q^7)G(q^{42}) - q^7b(q^7)H(q^{28})\}H(q^2)$$
$$- qG(q^2)\{q^7a(q^7)H(q^{42}) + b(q^7)G(q^{28})\} = \frac{\chi(-q)}{\chi(-q^7)}.$$

Upon rearrangement and the use of (5.14.2) and (3.11) with q replaced by q^2, we find that

$$a(q^7)N(q^2) - qb(q^7)\frac{\chi(-q^{14})}{\chi(-q^2)} = \frac{\chi(-q)}{\chi(-q^7)},$$

from which, by (4.38), we conclude that

$$(5.14.12) \qquad N(q^2) = \frac{1}{\chi^2(q^7)}\left\{\frac{\chi(-q)\chi(-q^{42})}{\chi(-q^7)\chi(-q^{14})} + q\frac{\chi(-q^7)\chi(-q^{14})}{\chi(-q^2)\chi(-q^{21})}\right\}.$$

Similarly, employing Lemma 4.4 in (3.11), we find that

$$\{a(q)G(q^6) - qb(q)H(q^4)\}G(q^{14})$$
$$+ q^3\{qa(q)H(q^6) + b(q)G(q^4)\}H(q^{14}) = \frac{\chi(-q^7)}{\chi(-q)}.$$

Upon rearrangement and the use of (5.14.1) and (3.10) with q replaced by q^2, we obtain

$$a(q)M(q^2) - qb(q)\frac{\chi(-q^2)}{\chi(-q^{14})} = \frac{\chi(-q^7)}{\chi(-q)},$$

from which, we similarly find that

$$(5.14.13) \qquad M(q^2) = \frac{1}{\chi^2(q)}\left\{\frac{\chi(-q^6)\chi(-q^7)}{\chi(-q)\chi(-q^2)} + q\frac{\chi(-q)\chi(-q^2)}{\chi(-q^3)\chi(-q^{14})}\right\}.$$

Next, recall that [5, p. 124, Entries 12 (v), (vi), (vii)]

$$(5.14.14) \quad \chi(q) = 2^{1/6}\left(\frac{q}{\alpha(1-\alpha)}\right)^{1/24} \quad \text{and} \quad \chi(-q) = 2^{1/6}\left(\frac{(1-\alpha)^2 q}{\alpha}\right)^{1/24}.$$

Let α, β, γ, and δ be of degrees 1, 3, 7, and 21, respectively. In (5.14.13), we use the representations (5.14.14) and (5.5.14) and conclude, after some algebra, that

$$M(q^2) = \frac{2^{-1/3}q^{1/3}}{\{\alpha\beta^2\gamma(1-\alpha)(1-\beta)^2(1-\gamma)\}^{1/24}}\{\alpha^{1/4}(1-\beta)^{1/8}(1-\gamma)^{1/8}$$
$$(5.14.15) \qquad\qquad\qquad\qquad\qquad + \beta^{1/8}\gamma^{1/8}(1-\alpha)^{1/4}\}.$$

Similarly, from (5.14.12), we find that

$$N(q^2) = \frac{2^{-1/3}q^{1/3}}{\{\alpha\gamma\delta^2(1-\alpha)(1-\gamma)(1-\delta)^2\}^{1/24}}\{\gamma^{1/4}(1-\alpha)^{1/8}(1-\delta)^{1/8}$$
$$(5.14.16) \qquad\qquad\qquad\qquad\qquad + \alpha^{1/8}\delta^{1/8}(1-\gamma)^{1/4}\}.$$

Lastly, from (5.14.11), we conclude that

$$J(q^2) = \frac{2^{-1/3} q^{1/3}}{\{\alpha\beta\gamma\delta(1-\alpha)(1-\beta)(1-\gamma)(1-\delta)\}^{1/24}} \{\{(1-\beta)(1-\delta)\}^{1/8}$$
(5.14.17)
$$- \{(1-\alpha)(1-\gamma)\}^{1/8}\}.$$

Therefore, the equation $M(q^2)N(q^2) = J^2(q^2)$ is equivalent to the modular equation
(5.14.18)
$$\left\{\{(1-\beta)(1-\delta)\}^{1/8} - \{(1-\alpha)(1-\gamma)\}^{1/8}\right\}^2$$
$$= \left\{\alpha^{1/4}(1-\beta)^{1/8}(1-\gamma)^{1/8} + \beta^{1/8}\gamma^{1/8}(1-\alpha)^{1/4}\right\}\left\{\gamma^{1/4}(1-\alpha)^{1/8}(1-\delta)^{1/8}\right.$$
$$\left. + \alpha^{1/8}\delta^{1/8}(1-\gamma)^{1/4}\right\}$$
$$= \left\{\alpha^2\gamma^2(1-\alpha)(1-\beta)(1-\gamma)(1-\delta)\right\}^{1/8} + \left\{\alpha^3\delta(1-\beta)(1-\gamma)^3\right\}^{1/8}$$
$$+ \left\{\gamma^3\beta(1-\delta)(1-\alpha)^3\right\}^{1/8} + \left\{\alpha\beta\gamma\delta(1-\alpha)^2(1-\gamma)^2\right\}^{1/8}.$$

To prove (5.14.18), we invoke two modular equations, of degrees 3 and 7, respectively. Namely, if β has degree 3 over α, then [**5**, p. 230, Entry 5 (i)]
(5.14.19)
$$\{\alpha^3(1-\beta)\}^{1/8} - \{\beta(1-\alpha)^3\}^{1/8} = \{\beta(1-\beta)\}^{1/8},$$
and, if γ has degree 7 over α, then [**5**, p. 314, Entry 19 (i)]
(5.14.20)
$$\{(1-\alpha)(1-\gamma)\}^{1/8} + \{\alpha\gamma\}^{1/8} = 1.$$

Let
$$u := (\alpha\gamma)^{1/8}, \quad v := (\beta\delta)^{1/8}, \quad x := \{\beta(1-\alpha)^3\}^{1/8},$$
$$y := \{\alpha^3(1-\beta)\}^{1/8}, \quad \bar{x} := \{\delta(1-\gamma)^3\}^{1/8}, \quad \text{and} \quad \bar{y} := \{\gamma^3(1-\delta)\}^{1/8}.$$

Since γ has degree 7 over α and δ has degree 7 over β, by (5.14.20),
$$\{(1-\alpha)(1-\gamma)\}^{1/8} = 1 - u \quad \text{and} \quad \{(1-\beta)(1-\delta)\}^{1/8} = 1 - v.$$
Since β has degree 3 over α and δ has degree 3 over γ, by (5.14.19),
$$y - x = \{\beta(1-\beta)\}^{1/8} \quad \text{and} \quad \bar{y} - \bar{x} = \{\delta(1-\delta)\}^{1/8}.$$
By using the trivial identity
$$y\bar{x} + \bar{y}x = x\bar{x} + y\bar{y} - (x-y)(\bar{x} - \bar{y}),$$
we conclude that
$$y\bar{x} + \bar{y}x = v(1-u)^3 + u^3(1-v) - v(1-v).$$
Returning to the equation (5.14.18), we see that the right-hand side of (5.14.18) is
$$u^2(1-u)(1-v) + y\bar{x} + \bar{y}x + uv(1-u)^2$$
$$= u^2(1-u)(1-v) + v(1-u)^3 + u^3(1-v) - v(1-v) + uv(1-u)^2,$$
which, after some algebra, simplifies to
$$(u-v)^2 = \{(1-v) - (1-u)\}^2,$$
which is exactly the far left side of (5.14.18). Hence, the proof of (3.17) is complete.

5.15. Proof of Entry 3.16

We prove that both sides of (3.18) are independently equal to the right-hand side of (3.19). For brevity of exposition, we make the following definition. Assuming that S is a subset of the rational numbers and $\sum_{n \in S} a_n q^n$ is a generic q-series, we define an operator $\mathcal{L}$ acting on $\sum_{n \in S} a_n q^n$ by $\mathcal{L}\left(\sum_{n \in S} a_n q^n\right) = \sum_{n \in S'} a_n q^n$, where $S' \subseteq S$ is the set of all integers in S.

We apply Lemma 5.1 with q replaced by q^2 and q^{13} to respectively deduce that

(5.15.1)
$$f(-q^2)f(-q^{2/5}) = f^2(-q^4,-q^6) - q^{4/5}f^2(-q^2,-q^8) - q^{2/5}f(-q^2)f(-q^{10}),$$

(5.15.2)
$$f(-q^{13})f(-q^{13/5}) = f^2(-q^{26},-q^{39}) - q^{26/5}f^2(-q^{13},-q^{52}) - q^{13/5}f(-q^{13})f(-q^{65}).$$

Multiplying together (5.15.1) and (5.15.2), we obtain

(5.15.3)
$$\begin{aligned} f(-q^2)f(-q^{13})&f(-q^{2/5})f(-q^{13/5}) = f^2(-q^4,-q^6)f^2(-q^{26},-q^{39}) \\ &+ q^6 f^2(-q^2,-q^8)f^2(-q^{13},-q^{52}) + q^3 f(-q^2)f(-q^{10})f(-q^{13})f(-q^{65}) \\ &- q^{26/5} f^2(-q^4,-q^6)f^2(-q^{13},-q^{52}) - q^{13/5} f^2(-q^4,-q^6)f(-q^{13})f(-q^{65}) \\ &- q^{4/5} f^2(-q^2,-q^8)f^2(-q^{26},-q^{39}) + q^{17/5} f^2(-q^2,-q^8)f(-q^{13})f(-q^{65}) \\ &- q^{2/5} f^2(-q^{26},-q^{39})f(-q^2)f(-q^{10}) + q^{28/5} f^2(-q^{13},-q^{52})f(-q^2)f(-q^{10}). \end{aligned}$$

We consider terms with integral powers of q on both sides of (5.15.3) and observe that

(5.15.4)
$$\begin{aligned} \mathcal{L}\left(f(-q^2)f(-q^{13})f(-q^{2/5})f(-q^{13/5})\right) \\ = f^2(-q^4,-q^6)f^2(-q^{26},-q^{39}) + q^6 f^2(-q^2,-q^8)f^2(-q^{13},-q^{52}) \\ + q^3 f(-q^2)f(-q^{10})f(-q^{13})f(-q^{65}). \end{aligned}$$

We now derive an alternative expression for the left-hand side of (5.15.4) above. To this end, we first employ (2.19) with $a = -q$, $b = -q^2$, and $n = 5$ to deduce that

(5.15.5)
$$\begin{aligned} f(-q) &= f(-q^{35},-q^{40}) - qf(-q^{50},-q^{25}) + q^5 f(-q^{65},-q^{10}) - q^{12} f(-q^{80},-q^{-5}) \\ &\quad + q^{22} f(-q^{95},-q^{-20}) \\ &= f(-q^{35},-q^{40}) - qf(-q^{50},-q^{25}) + q^5 f(-q^{65},-q^{10}) + q^7 f(-q^{70},-q^5) \\ &\quad - q^2 f(-q^{20},-q^{55}), \end{aligned}$$

after two applications of (2.5). We then apply (5.15.5) above to obtain representations for $f(-q^{2/5})$ and $f(-q^{13/5})$ by replacing q by $q^{2/5}$ and q by $q^{13/5}$, respectively. This gives us

(5.15.6)
$$\begin{aligned} f(-q^{2/5}) = f(-q^{14},-q^{16}) - q^{2/5} f(-q^{20},-q^{10}) + q^2 f(-q^{26},-q^4) \\ + q^{14/5} f(-q^{28},-q^2) - q^{4/5} f(-q^8,-q^{22}) \end{aligned}$$

and
$$f(-q^{13/5}) = f(-q^{91}, -q^{104}) - q^{13/5} f(-q^{130}, -q^{65}) + q^{13} f(-q^{169}, -q^{26})$$
(5.15.7)
$$+ q^{91/5} f(-q^{182}, -q^{13}) - q^{26/5} f(-q^{52}, -q^{143}).$$

Thus, multiplying (5.15.6) and (5.15.7), we see that
$$\mathcal{L}\left(f(-q^{2/5}) f(-q^{13/5})\right) = f(-q^{14}, -q^{16}) f(-q^{91}, -q^{104})$$
$$+ q^3 f(-q^{10}, -q^{20}) f(-q^{65}, -q^{130}) + q^{15} f(-q^4, -q^{26}) f(-q^{26}, -q^{169})$$
$$+ q^{21} f(-q^2, -q^{28}) f(-q^{13}, -q^{182}) + q^6 f(-q^8, -q^{22}) f(-q^{52}, -q^{143})$$
$$+ q^{13} f(-q^{14}, -q^{16}) f(-q^{26}, -q^{169}) + q^2 f(-q^4, -q^{26}) f(-q^{91}, -q^{104})$$
(5.15.8)
$$- q^8 f(-q^2, -q^{28}) f(-q^{52}, -q^{143}) - q^{19} f(-q^8, -q^{22}) f(-q^{13}, -q^{182}).$$

Since $f(-q^2) f(-q^{13})$ contains only integral powers of q, it follows that
(5.15.9)
$$\mathcal{L}\left(f(-q^2) f(-q^{13}) f(-q^{2/5}) f(-q^{13/5})\right) = f(-q^2) f(-q^{13}) \mathcal{L}\left(f(-q^{2/5}) f(-q^{13/5})\right)$$
$$= f(-q^2) f(-q^{13}) \{ f(-q^{14}, -q^{16}) f(-q^{91}, -q^{104}) + q^3 f(-q^{10}, -q^{20}) f(-q^{65}, -q^{130})$$
$$+ q^{15} f(-q^4, -q^{26}) f(-q^{26}, -q^{169}) + q^{21} f(-q^2, -q^{28}) f(-q^{13}, -q^{182})$$
$$+ q^6 f(-q^8, -q^{22}) f(-q^{52}, -q^{143}) + q^{13} f(-q^{14}, -q^{16}) f(-q^{26}, -q^{169})$$
$$+ q^2 f(-q^4, -q^{26}) f(-q^{91}, -q^{104}) - q^8 f(-q^2, -q^{28}) f(-q^{52}, -q^{143})$$
$$- q^{19} f(-q^8, -q^{22}) f(-q^{13}, -q^{182}) \}.$$

Equating the right-hand sides of (5.15.4) and (5.15.9), we deduce that
(5.15.10)
$$f^2(-q^4, -q^6) f^2(-q^{26}, -q^{39}) + q^6 f^2(-q^2, -q^8) f^2(-q^{13}, -q^{52})$$
$$+ q^3 f(-q^2) f(-q^{10}) f(-q^{13}) f(-q^{65})$$
$$= f(-q^2) f(-q^{13}) \{ f(-q^{14}, -q^{16}) f(-q^{91}, -q^{104}) + q^3 f(-q^{10}, -q^{20}) f(-q^{65}, -q^{130})$$
$$+ q^{15} f(-q^4, -q^{26}) f(-q^{26}, -q^{169}) + q^{21} f(-q^2, -q^{28}) f(-q^{13}, -q^{182})$$
$$+ q^6 f(-q^8, -q^{22}) f(-q^{52}, -q^{143}) + q^{13} f(-q^{14}, -q^{16}) f(-q^{26}, -q^{169})$$
$$+ q^2 f(-q^4, -q^{26}) f(-q^{91}, -q^{104}) - q^8 f(-q^2, -q^{28}) f(-q^{52}, -q^{143})$$
$$- q^{19} f(-q^8, -q^{22}) f(-q^{13}, -q^{182}) \}.$$

We seek to simplify the right-hand side of (5.15.10). Applying (4.13) with $\alpha = 1$, $\beta = \frac{13}{2}$, $m = 1$, and $p = 15$, we see that

$$q^{-1/8} \sum_{k=1}^{7} F(1, \tfrac{13}{2}, 1, 15, \tfrac{1}{2}, k) = f(-q^{14}, -q^{16}) f(-q^{91}, -q^{104})$$
$$+ q f(-q^{12}, -q^{18}) f(-q^{78}, -q^{117}) + q^3 f(-q^{10}, -q^{20}) f(-q^{65}, -q^{130})$$
$$+ q^6 f(-q^8, -q^{22}) f(-q^{52}, -q^{143}) + q^{10} f(-q^6, -q^{24}) f(-q^{39}, -q^{156})$$
(5.15.11)
$$+ q^{15} f(-q^4, -q^{26}) f(-q^{26}, -q^{169}) + q^{21} f(-q^2, -q^{28}) f(-q^{13}, -q^{182}).$$

We now observe that five out of the seven terms appearing on the right-hand side of (5.15.11) also appear on the right-hand side of (5.15.10). This enables us to rewrite

(5.15.10) as

(5.15.12)
$$\begin{aligned}
&f^2(-q^4,-q^6)f^2(-q^{26},-q^{39}) + q^6 f^2(-q^2,-q^8)f^2(-q^{13},-q^{52}) \\
&\quad + q^3 f(-q^2)f(-q^{10})f(-q^{13})f(-q^{65}) \\
&= f(-q^2)f(-q^{13})\Big\{q^{-1/8}\sum_{k=1}^{7} F(1,\tfrac{13}{2},1,15,\tfrac{1}{2},k) - qf(-q^{12},-q^{18})f(-q^{78},-q^{117}) \\
&\quad - q^{10} f(-q^6,-q^{24})f(-q^{39},-q^{156}) + q^{13} f(-q^{14},-q^{16})f(-q^{26},-q^{169}) \\
&\quad + q^2 f(-q^4,-q^{26})f(-q^{91},-q^{104}) - q^8 f(-q^2,-q^{28})f(-q^{52},-q^{143}) \\
&\quad - q^{19} f(-q^8,-q^{22})f(-q^{13},-q^{182})\Big\}.
\end{aligned}$$

We next apply (4.13) again with $\alpha=1$, $\beta=13/2$, $m=11$, and $p=15$. This yields

(5.15.13)
$$\begin{aligned}
q^{-1/8}\sum_{k=1}^{7} F(1,\tfrac{13}{2},11,15,\tfrac{17}{2},k) &= q^2 f(-q^{26},-q^4)f(-q^{104},-q^{91}) \\
&\quad + q^{19} f(-q^{48},-q^{-18})f(-q^{117},-q^{78}) + q^{53} f(-q^{70},-q^{-40})f(-q^{130},-q^{65}) \\
&\quad + q^{104} f(-q^{92},-q^{-62})f(-q^{143},-q^{52}) + q^{172} f(-q^{114},-q^{-84})f(-q^{156},-q^{39}) \\
&\quad + q^{257} f(-q^{136},-q^{-106})f(-q^{169},-q^{26}) + q^{359} f(-q^{158},-q^{-128})f(-q^{182},-q^{13}).
\end{aligned}$$

After several applications of (2.5), we rewrite (5.15.13) as

$$\begin{aligned}
q^{-1/8}\sum_{k=1}^{7} F(1,\tfrac{13}{2},11,15,\tfrac{17}{2},k) &= q^2 f(-q^4,-q^{26})f(-q^{91},-q^{104}) \\
&\quad - qf(-q^{12},-q^{18})f(-q^{78},-q^{117}) + q^3 f(-q^{10},-q^{20})f(-q^{65},-q^{130}) \\
&\quad - q^8 f(-q^2,-q^{28})f(-q^{52},-q^{143}) - q^{10} f(-q^6,-q^{24})f(-q^{39},-q^{156})
\end{aligned}$$
(5.15.14) $\qquad + q^{13} f(-q^{14},-q^{16})f(-q^{26},-q^{169}) - q^{19} f(-q^8,-q^{22})f(-q^{13},-q^{182}).$

We now note from (2.9) that $q^3 f(-q^{10},-q^{20})f(-q^{65},-q^{130}) = q^3 f(-q^{10})f(-q^{65})$, and upon comparing the right-hand side of (5.15.14) with that of (5.15.12), we rewrite (5.15.12) as

(5.15.15)
$$\begin{aligned}
&f^2(-q^4,-q^6)f^2(-q^{26},-q^{39}) + q^6 f^2(-q^2,-q^8)f^2(-q^{13},-q^{52}) \\
&\quad + q^3 f(-q^2)f(-q^{10})f(-q^{13})f(-q^{65}) \\
&= f(-q^2)f(-q^{13})\Big\{q^{-1/8}\sum_{k=1}^{7} F(1,\tfrac{13}{2},1,15,\tfrac{1}{2},k) + q^{-1/8}\sum_{k=1}^{7} F(1,\tfrac{13}{2},11,15,\tfrac{17}{2},k) \\
&\quad - q^3 f(-q^{10})f(-q^{65})\Big\},
\end{aligned}$$

or equivalently upon applying the Jacobi triple product identity (2.6) to $f(-q^4, -q^6)$, $f(-q^{26}, -q^{39})$, $f(-q^2, -q^8)$, and $f(-q^{13}, -q^{52})$, we deduce that

(5.15.16)
$$(f(-q^4, -q^6)f(-q^{26}, -q^{39}) + q^3 f(-q^2, -q^8)f(-q^{13}, -q^{52}))^2$$
$$= f(-q^2)f(-q^{13})\left\{q^{-1/8}\sum_{k=1}^{7} F(1, \tfrac{13}{2}, 1, 15, \tfrac{1}{2}, k) + q^{-1/8}\sum_{k=1}^{7} F(1, \tfrac{13}{2}, 11, 15, \tfrac{17}{2}, k)\right\}.$$

We now turn our attention to the two sums appearing on the right-hand side of (5.15.16). From (4.13), we see that

(5.15.17)
$$q^{-1/8}\sum_{k=1}^{7} F(1, \tfrac{13}{2}, 1, 15, \tfrac{1}{2}, k) = \frac{q^{-1/8}}{2}\sum_{u=-\infty}^{\infty}\sum_{t=-\infty}^{\infty}(-1)^t q^{\frac{1}{2}\{u+\frac{1}{2}+2t\}^2 + 13t^2}$$
$$= \frac{q^{-1/8}}{2}\sum_{u=-\infty}^{\infty}\sum_{t=-\infty}^{\infty}(-1)^t q^{\frac{1}{2}\{u+\frac{1}{2}\}^2 + 13t^2} = \frac{q^{-1/8}}{2}\sum_{u=-\infty}^{\infty} q^{\frac{1}{2}\{u+\frac{1}{2}\}^2}\sum_{t=-\infty}^{\infty}(-1)^t q^{13t^2}$$
$$= \psi(q)\varphi(-q^{13}) = \frac{(q^2; q^2)_\infty}{(q; q^2)_\infty}(q^{13}; q^{26})_\infty^2(q^{26}; q^{26})_\infty = \frac{(q^2; q^2)_\infty^2 (q^{13}; q^{13})_\infty^2}{(q; q)_\infty (q^{26}; q^{26})_\infty}$$
$$= \frac{f^2(-q^2)f^2(-q^{13})}{f(-q)f(-q^{26})},$$

where we have utilized (2.7)–(2.9). Similarly, from (4.13), we find that

(5.15.18)
$$q^{-1/8}\sum_{k=1}^{7} F(1, \tfrac{13}{2}, 11, 15, \tfrac{17}{2}, k) = \frac{q^{-1/8}}{2}\sum_{u=-\infty}^{\infty}\sum_{t=-\infty}^{\infty}(-1)^t q^{\frac{17}{2}\{u+\frac{1}{2}+\frac{22}{17}t\}^2 + \frac{13}{17}t^2}$$
$$= \frac{q^{-1/8}}{2}\sum_{u=-\infty}^{\infty}\sum_{t=-\infty}^{\infty}(-1)^t q^{\frac{17}{2}\{u+\frac{1}{2}+\frac{5}{17}t\}^2 + \frac{13}{17}t^2}$$
$$= \frac{1}{2}\sum_{u=-\infty}^{\infty}\sum_{t=-\infty}^{\infty}(-1)^t q^{2+\frac{3}{2}t^2+\frac{5}{2}t+5tu+\frac{17}{2}u(u+1)}$$
$$= -\frac{q}{2}\sum_{u=-\infty}^{\infty}\sum_{t=-\infty}^{\infty}(-1)^t q^{\frac{3}{2}t^2-\frac{1}{2}t+5tu+\frac{17}{2}u^2+\frac{7}{2}u},$$

where in the last equality we replaced t by $t-1$.

We now claim that

(5.15.19) $$\frac{1}{2}\sum_{u=-\infty}^{\infty}\sum_{t=-\infty}^{\infty}(-1)^t q^{\frac{3}{2}t^2-\frac{1}{2}t+5tu+\frac{17}{2}u^2+\frac{7}{2}u} = f(-q)f(-q^{26}).$$

To this end, we dissect the series according as $u \equiv 0, 1, -1 \pmod{3}$ respectively. We consider each of the three sums separately. If we replace u by $3u$ and t by $-t - 5u$, we find that

$$\sum_{\substack{u=-\infty \\ u \equiv 0 \,(\mathrm{mod}\, 3)}}^{\infty} \sum_{t=-\infty}^{\infty} (-1)^t q^{\frac{3}{2}t^2 - \frac{1}{2}t + 5tu + \frac{17}{2}u^2 + \frac{7}{2}u}$$

$$= \sum_{u=-\infty}^{\infty} \sum_{t=-\infty}^{\infty} (-1)^{t+u} q^{\frac{3}{2}t^2 + \frac{1}{2}t + 39u^2 + 13u}$$

(5.15.20)
$$= \sum_{u=-\infty}^{\infty} (-1)^u q^{39u^2 + 13u} \sum_{t=-\infty}^{\infty} (-1)^t q^{\frac{3}{2}t^2 + \frac{1}{2}t} = f(-q^{26}) f(-q),$$

by (2.9). Next, if we replace u by $3u+1$ and t by $-t-5u$ in the series in (5.15.19), we find that

$$\sum_{\substack{u=-\infty \\ u \equiv 1 \,(\mathrm{mod}\, 3)}}^{\infty} \sum_{t=-\infty}^{\infty} (-1)^t q^{\frac{3}{2}t^2 - \frac{1}{2}t + 5tu + \frac{17}{2}u^2 + \frac{7}{2}u}$$

$$= \sum_{u=-\infty}^{\infty} \sum_{t=-\infty}^{\infty} (-1)^{t+u} q^{12 + \frac{3}{2}t^2 - \frac{9}{2}t + 39u^2 + 39u}$$

(5.15.21)
$$= \sum_{u=-\infty}^{\infty} (-1)^u q^{39u^2 + 39u} \sum_{t=-\infty}^{\infty} (-1)^t q^{12 + \frac{3}{2}t^2 - \frac{9}{2}t} = 0,$$

by (2.4). Finally, if we replace u by $3u-1$ and t by $-t-5u+2$ in the series in (5.15.19), we see that

$$\sum_{\substack{u=-\infty \\ u \equiv -1 \,(\mathrm{mod}\, 3)}}^{\infty} \sum_{t=-\infty}^{\infty} (-1)^t q^{\frac{3}{2}t^2 - \frac{1}{2}t + 5tu + \frac{17}{2}u^2 + \frac{7}{2}u}$$

$$= \sum_{u=-\infty}^{\infty} \sum_{t=-\infty}^{\infty} (-1)^{t+u} q^{\frac{3}{2}t^2 - \frac{1}{2}t + 39u^2 - 13u}$$

(5.15.22)
$$= \sum_{u=-\infty}^{\infty} (-1)^u q^{39u^2 - 13u} \sum_{t=-\infty}^{\infty} (-1)^t q^{\frac{3}{2}t^2 - \frac{1}{2}t} = f(-q^{26}) f(-q),$$

by (2.9). Thus, in (5.15.20)–(5.15.22), we have shown that

(5.15.23)
$$\frac{1}{2} \sum_{u=-\infty}^{\infty} \sum_{t=-\infty}^{\infty} (-1)^t q^{\frac{3}{2}t^2 - \frac{1}{2}t + 5tu + \frac{17}{2}u^2 + \frac{7}{2}u} = \frac{1}{2} (f(-q) f(-q^{26}) + f(-q) f(-q^{26}))$$
$$= f(-q) f(-q^{26}),$$

as claimed in (5.15.19). Thus, from (5.15.18) and (5.15.23), we deduce that

(5.15.24)
$$q^{-1/8} \sum_{k=1}^{7} F(1, \tfrac{13}{2}, 11, 15, \tfrac{17}{2}, k) = -q f(-q) f(-q^{26}).$$

Finally, we insert (5.15.17) and (5.15.24) into (5.15.16) to arrive at

$$(f(-q^4,-q^6)f(-q^{26},-q^{39}) + q^3 f(-q^2,-q^8)f(-q^{13},-q^{52}))^2$$
(5.15.25)
$$= f(-q^2)f(-q^{13})\left\{\frac{f^2(-q^2)f^2(-q^{13})}{f(-q)f(-q^{26})} - qf(-q)f(-q^{26})\right\}.$$

Dividing both sides of (5.15.25) by $f^2(-q^2)f^2(-q^{13})$ and taking square roots, we obtain

$$\frac{f(-q^4,-q^6)f(-q^{26},-q^{39})}{f(-q^2)f(-q^{13})} + q^3 \frac{f(-q^2,-q^8)f(-q^{13},-q^{52})}{f(-q^2)f(-q^{13})}$$

$$= \sqrt{\frac{f(-q^2)f(-q^{13})}{f(-q)f(-q^{26})} - q\frac{f(-q)f(-q^{26})}{f(-q^2)f(-q^{13})}}$$

(5.15.26)
$$= \sqrt{\frac{\chi(-q^{13})}{\chi(-q)} - q\frac{\chi(-q)}{\chi(-q^{13})}},$$

by (2.14). Using (2.11), we see that we have shown that the left side in (3.18) is equal to (3.19).

We now turn to the right-hand side of (3.18) and show that it equals the expression in (3.19). Our argument is brief, since the proof is similar to the previous proof above. We apply Lemma 5.1, and then apply it a second time with q replaced by q^{26}. Then multiply the two resulting equalities together to obtain

(5.15.27)
$$f(-q)f(-q^{26})f(-q^{1/5})f(-q^{26/5}) = f^2(-q^2,-q^3)f^2(-q^{52},-q^{78})$$
$$- q^{2/5}f^2(-q,-q^4)f^2(-q^{52},-q^{78}) - q^{1/5}f(-q)f(-q^5)f^2(-q^{52},-q^{78})$$
$$- q^{52/5}f^2(-q^2,-q^3)f^2(-q^{26},-q^{104}) + q^{54/5}f^2(-q,-q^4)f^2(-q^{26},-q^{104})$$
$$+ q^{53/5}f(-q)f(-q^5)f^2(-q^{26},-q^{104}) - q^{26/5}f^2(-q^2,-q^3)f(-q^{26})f(-q^{130})$$
$$+ q^{28/5}f^2(-q,-q^4)f(-q^{26})f(-q^{130}) + q^{27/5}f(-q)f(-q^5)f(-q^{26})f(-q^{130}).$$

Recalling the definition of $\mathcal{L}$ at the beginning of this section, we have

(5.15.28)
$$\mathcal{L}\left(q^{-2/5}f(-q)f(-q^{26})f(-q^{1/5})f(-q^{26/5})\right) = -f^2(-q,-q^4)f^2(-q^{52},-q^{78})$$
$$- q^{10}f^2(-q^2,-q^3)f^2(-q^{26},-q^{104}) + q^5 f(-q)f(-q^5)f(-q^{26})f(-q^{130}).$$

We then apply (5.15.5) to obtain representations for $f(-q^{1/5})$ and $f(-q^{26/5})$ by replacing q by $q^{1/5}$ and q by $q^{26/5}$, respectively. This gives us

(5.15.29)
$$f(-q^{1/5}) = f(-q^7,-q^8) - q^{1/5}f(-q^{10},-q^5) + qf(-q^{13},-q^2)$$
$$+ q^{7/5}f(-q^{14},-q) - q^{2/5}f(-q^4,-q^{11})$$

and

(5.15.30)
$$f(-q^{26/5}) = f(-q^{182},-q^{208}) - q^{26/5}f(-q^{260},-q^{130}) + q^{26}f(-q^{338},-q^{52})$$
$$+ q^{182/5}f(-q^{364},-q^{26}) - q^{52/5}f(-q^{104},-q^{286}).$$

Thus, since $f(-q)f(-q^{26})$ contains only terms with integral powers, we find upon using (5.15.29) and (5.15.30) that

(5.15.31)
$$\mathcal{L}\left(q^{-2/5}f(-q)f(-q^{26})f(-q^{1/5})f(-q^{26/5})\right)$$
$$= f(-q)f(-q^{26})\mathcal{L}\left(q^{-2/5}f(-q^{1/5})f(-q^{26/5})\right)$$
$$= f(-q)f(-q^{26})\{-f(-q^4,-q^{11})f(-q^{182},-q^{208}) + q^{37}f(-q^2,-q^{13})f(-q^{26},-q^{364})$$
$$+ q^{36}f(-q^7,-q^8)f(-q^{26},-q^{364}) + q^5 f(-q^5,-q^{10})f(-q^{130},-q^{260})$$
$$- q^{10}f(-q^7,-q^8)f(-q^{104},-q^{286}) - q^{11}f(-q^2,-q^{13})f(-q^{104},-q^{286})$$
$$+ qf(-q,-q^{14})f(-q^{182},-q^{208}) + q^{27}f(-q,-q^{14})f(-q^{52},-q^{338})$$
$$- q^{26}f(-q^4,-q^{11})f(-q^{52},-q^{338})\}.$$

Now from (4.13) and several applications of (2.5), we see that

$$q^{-5/8}\sum_{k=1}^{7} F(\tfrac{1}{2},13,7,15,\tfrac{5}{2},k) = f(-q^4,-q^{11})f(-q^{182},-q^{208})$$
$$- q^2 f(-q^3,-q^{12})f(-q^{156},-q^{234}) - q^5 f(-q^5,-q^{10})f(-q^{130},-q^{260})$$
$$+ q^{11}f(-q^2,-q^{13})f(-q^{104},-q^{286}) + q^{17}f(-q^6,-q^9)f(-q^{78},-q^{312})$$
(5.15.32) $$- q^{27}f(-q,-q^{14})f(-q^{52},-q^{338}) - q^{36}f(-q^7,-q^8)f(-q^{26},-q^{364})$$

and

$$q^{-5/8}\sum_{k=1}^{7} F(\tfrac{1}{2},13,13,15,\tfrac{13}{2},k) = qf(-q,-q^{14})f(-q^{182},-q^{208})$$
$$- q^2 f(-q^3,-q^{12})f(-q^{156},-q^{234}) + q^5 f(-q^5,-q^{10})f(-q^{130},-q^{260})$$
$$- q^{10}f(-q^7,-q^8)f(-q^{104},-q^{286}) + q^{17}f(-q^6,-q^9)f(-q^{78},-q^{312})$$
(5.15.33) $$- q^{26}f(-q^4,-q^{11})f(-q^{52},-q^{338}) + q^{37}f(-q^2,-q^{13})f(-q^{26},-q^{364}).$$

Comparing the right-hand sides of (5.15.31), (5.15.32), and (5.15.33), we see that

$$\mathcal{L}\left(q^{-2/5}f(-q)f(-q^{26})f(-q^{1/5})f(-q^{26/5})\right)$$
$$= f(-q)f(-q^{26})\Big\{-q^{-5/8}\sum_{k=1}^{7} F(\tfrac{1}{2},13,7,15,\tfrac{5}{2},k)$$
(5.15.34) $$+ q^{-5/8}\sum_{k=1}^{7} F(\tfrac{1}{2},13,13,15,\tfrac{13}{2},k) - q^5 f(-q^5)f(-q^{130})\Big\}.$$

Now, combining the right-hand sides of (5.15.28) and (5.15.34), applying the Jacobi triple product identity (2.6) to $f(-q,-q^4)$, $f(-q^{52},-q^{78})$, $f(-q^2,-q^3)$, and

$f(-q^{26}, -q^{104})$, and simplifying, we deduce that

$$\left(f(-q, -q^4)f(-q^{52}, -q^{78}) - q^5 f(-q^2, -q^3)f(-q^{26}, -q^{104})\right)^2$$
$$= -f(-q)f(-q^{26})\left\{-q^{-5/8}\sum_{k=1}^{7} F(\tfrac{1}{2}, 13, 7, 15, \tfrac{5}{2}, k)\right.$$
(5.15.35)
$$\left. + q^{-5/8}\sum_{k=1}^{7} F(\tfrac{1}{2}, 13, 13, 15, \tfrac{13}{2}, k)\right\}.$$

We now concentrate on the two sums arising on the right-hand side of (5.15.35) above. From (4.13), we have

(5.15.36)
$$q^{-5/8}\sum_{k=1}^{7} F(\tfrac{1}{2}, 13, 13, 15, \tfrac{13}{2}, k) = \frac{q^{-5/8}}{2}\sum_{u=-\infty}^{\infty}\sum_{t=-\infty}^{\infty} (-1)^t q^{\frac{13}{2}\left\{u+\frac{1}{2}+t\right\}^2 + t^2}$$
$$= \frac{q^{-5/8}}{2}\sum_{u=-\infty}^{\infty}\sum_{t=-\infty}^{\infty}(-1)^t q^{\frac{13}{2}\left\{u+\frac{1}{2}\right\}^2+t^2} = \frac{q}{2}\sum_{u=-\infty}^{\infty} q^{\frac{13}{2}u(u+1)}\sum_{t=-\infty}^{\infty}(-1)^t q^{t^2}$$
$$= q\psi(q^{13})\varphi(-q) = q\frac{(q^{26}; q^{26})_\infty}{(q^{13}; q^{26})_\infty}(q; q^2)_\infty^2 (q^2; q^2)_\infty = q\frac{(q^{26}; q^{26})_\infty^2 (q; q)_\infty}{(q^{13}; q^{13})_\infty (q^2; q^2)_\infty}$$
$$= q\frac{f^2(-q^{26})f^2(-q)}{f(-q^{13})f(-q^2)},$$

where we have used (2.7)–(2.9). Similarly, from (4.13), we find that

(5.15.37)
$$q^{-5/8}\sum_{k=1}^{7} F(\tfrac{1}{2}, 13, 7, 15, \tfrac{5}{2}, k) = \frac{q^{-5/8}}{2}\sum_{u=-\infty}^{\infty}\sum_{t=-\infty}^{\infty}(-1)^t q^{\frac{5}{2}\left\{u+\frac{1}{2}+\frac{7}{5}t\right\}^2 + \frac{13}{5}t^2}$$
$$= \frac{q^{-5/8}}{2}\sum_{u=-\infty}^{\infty}\sum_{t=-\infty}^{\infty}(-1)^t q^{\frac{5}{2}\left\{u+\frac{1}{2}+\frac{2}{5}t\right\}^2 + \frac{13}{5}t^2}$$
$$= \frac{1}{2}\sum_{u=-\infty}^{\infty}\sum_{t=-\infty}^{\infty}(-1)^t q^{3t^2 + t + 2tu + \frac{5}{2}u(u+1)}.$$

As in the previous proof, we dissect the series above according as $u \equiv 0, 1, -1 \pmod{3}$. Assuming that $u \equiv 0 \pmod{3}$, we replace u by $3u$ and t by $-t - u$ to find that

(5.15.38)
$$\sum_{\substack{u=-\infty \\ u \equiv 0 \,(\mathrm{mod}\, 3)}}^{\infty}\sum_{t=-\infty}^{\infty}(-1)^t q^{3t^2 + t + 2tu + \frac{5}{2}u(u+1)} = \sum_{u=-\infty}^{\infty}\sum_{t=-\infty}^{\infty}(-1)^{t+u} q^{3t^2 - t + \frac{39}{2}u^2 + \frac{13}{2}u}$$
$$= \sum_{u=-\infty}^{\infty}(-1)^u q^{\frac{13}{2}u(3u+1)}\sum_{t=-\infty}^{\infty}(-1)^t q^{t(3t-1)} = f(-q^{13})f(-q^2),$$

by (2.9). Next, if we replace u by $3u+1$ and t by $t-u$ in the series in (5.15.19), we find that

$$\sum_{\substack{u=-\infty \\ u \equiv 1 \,(\mathrm{mod}\, 3)}}^{\infty} \sum_{t=-\infty}^{\infty} (-1)^t q^{3t^2+t+2tu+\frac{5}{2}u(u+1)}$$

$$= \sum_{u=-\infty}^{\infty} \sum_{t=-\infty}^{\infty} (-1)^{t+u} q^{3t^2+3t+\frac{39}{2}u^2+\frac{39}{2}u+5}$$

(5.15.39)
$$= \sum_{u=-\infty}^{\infty} (-1)^u q^{\frac{39}{2}u^2+\frac{39}{2}u+5} \sum_{t=-\infty}^{\infty} (-1)^t q^{3t^2+3t} = 0,$$

by (2.4). Finally, if we replace u by $3u-1$ and t by $t-u$, we obtain

$$\sum_{\substack{u=-\infty \\ u \equiv -1 \,(\mathrm{mod}\, 3)}}^{\infty} \sum_{t=-\infty}^{\infty} (-1)^t q^{3t^2+t+2tu+\frac{5}{2}u(u+1)}$$

$$= \sum_{u=-\infty}^{\infty} \sum_{t=-\infty}^{\infty} (-1)^{t+u} q^{3t^2-t+\frac{39}{2}u^2-\frac{13}{2}u}$$

(5.15.40)
$$= f(-q^{13})f(-q^2),$$

by (2.9). Thus, from (5.15.37)–(5.15.40), we have shown that

$$q^{-5/8} \sum_{k=1}^{7} F(\tfrac{1}{2}, 13, 7, 15, \tfrac{5}{2}, k) = \frac{1}{2}\left(f(-q^{13})f(-q^2)+f(-q^{13})f(-q^2)\right)$$

(5.15.41) $= f(-q^{13})f(-q^2).$

Finally, we insert (5.15.36) and (5.15.41) into (5.15.35) and conclude that

$$\left(f(-q,-q^4)f(-q^{52},-q^{78}) - q^5 f(-q^2,-q^3)f(-q^{26},-q^{104})\right)^2$$

(5.15.42)
$$= -f(-q)f(-q^{26})\left\{-f(-q^{13})f(-q^2) + q\frac{f^2(-q^{26})f^2(-q)}{f(-q^{13})f(-q^2)}\right\}.$$

Dividing both sides of (5.15.42) by $f^2(-q)f^2(-q^{26})$ and taking square roots, we arrive at

$$\frac{f(-q,-q^4)f(-q^{52},-q^{78}) - q^5 f(-q^2,-q^3)f(-q^{26},-q^{104})}{f(-q)f(-q^{26})}$$

$$= \sqrt{\frac{f(-q^2)f(-q^{13})}{f(-q)f(-q^{26})} - q\frac{f(-q)f(-q^{26})}{f(-q^2)f(-q^{13})}}$$

(5.15.43)
$$= \sqrt{\frac{\chi(-q^{13})}{\chi(-q)} - q\frac{\chi(-q)}{\chi(-q^{13})}},$$

which completes the proof of the second part of Entry 3.16.

5.16. Proof of Entry 3.17

Using the product representations of $\chi(q)$ and $f(-q)$, given in (2.9) and (2.10), respectively, together with (2.7) and (2.8), we find that

(5.16.1) $\qquad \varphi(q) = f(q,q) = (-q;q^2)_\infty^2 (q^2;q^2)_\infty = \chi^2(q)f(-q^2)$

and

(5.16.2) $$\psi(q) = f(q, q^3) = \frac{(q^2; q^2)_\infty}{(q; q^2)_\infty} = \frac{(q; q)_\infty}{(q; q^2)_\infty} \frac{1}{(q; q^2)_\infty} = \frac{f(-q)}{\chi^2(-q)}.$$

By (2.11), (5.16.2), and (5.16.1) with q replaced by $q^{1/2}$ and $-q^{1/2}$, respectively, we find that Entry 3.17 is equivalent to the identity

(5.16.3)

$$f(-q^2, -q^3)f(-q^{38}, -q^{57}) + q^4 f(-q, -q^4)f(-q^{19}, -q^{76})$$
$$= f(-q)f(-q^{19}) \left\{ \frac{\chi^2(q^{1/2})\chi^2(q^{19/2})}{4\sqrt{q}} - \frac{\chi^2(-q^{1/2})\chi^2(-q^{19/2})}{4\sqrt{q}} - \frac{q^2}{\chi^2(-q)\chi^2(-q^{19})} \right\}$$
$$= \frac{1}{4\sqrt{q}} \left(\varphi(q^{1/2})\varphi(q^{19/2}) - \varphi(-q^{1/2})\varphi(-q^{19/2}) \right) - q^2 \psi(q)\psi(q^{19}).$$

We now apply Theorem 4.1 with the parameters $\epsilon_1 = \epsilon_2 = 0$, $a = b = q$, $c = d = q^{19}$, $\alpha = 1$, $\beta = 19$, and $m = 20$. Accordingly, we deduce that

(5.16.4)
$$\varphi(q)\varphi(q^{19}) = f(q^{20}, q^{20})f(q^{380}, q^{380}) + q^{19}f(q^{-18}, q^{58})f(q^{342}, q^{418})$$
$$+ q^{76}f(q^{-56}, q^{96})f(q^{304}, q^{456}) + q^{171}f(q^{-94}, q^{134})f(q^{266}, q^{494})$$
$$+ q^{304}f(q^{-132}, q^{172})f(q^{228}, q^{532}) + q^{475}f(q^{-170}, q^{210})f(q^{190}, q^{570})$$
$$+ q^{684}f(q^{-208}, q^{248})f(q^{152}, q^{608}) + q^{931}f(q^{-246}, q^{286})f(q^{114}, q^{646})$$
$$+ q^{1216}f(q^{-284}, q^{324})f(q^{76}, q^{684}) + q^{1539}f(q^{-322}, q^{362})f(q^{38}, q^{722})$$
$$+ q^{1900}f(q^{-360}, q^{400})f(1, q^{760}) + q^{2299}f(q^{-398}, q^{438})f(q^{-38}, q^{798})$$
$$+ q^{2736}f(q^{-436}, q^{476})f(q^{-76}, q^{836}) + q^{3211}f(q^{-474}, q^{514})f(q^{-114}, q^{874})$$
$$+ q^{3724}f(q^{-512}, q^{552})f(q^{-152}, q^{912}) + q^{4275}f(q^{-550}, q^{590})f(q^{-190}, q^{950})$$
$$+ q^{4864}f(q^{-588}, q^{628})f(q^{-228}, q^{988}) + q^{5491}f(q^{-626}, q^{666})f(q^{-266}, q^{1026})$$
$$+ q^{6156}f(q^{-664}, q^{704})f(q^{-304}, q^{1064}) + q^{6859}f(q^{-702}, q^{742})f(q^{-342}, q^{1102})$$
$$= f(q^{20}, q^{20})f(q^{380}, q^{380}) + 2qf(q^{18}, q^{22})f(q^{342}, q^{418})$$
$$+ 2q^4 f(q^{16}, q^{24})f(q^{304}, q^{456}) + 2q^9 f(q^{14}, q^{26})f(q^{266}, q^{494})$$
$$+ 2q^{16} f(q^{12}, q^{28})f(q^{228}, q^{532}) + 2q^{25} f(q^{10}, q^{30})f(q^{190}, q^{570})$$
$$+ 2q^{36} f(q^8, q^{32})f(q^{152}, q^{608}) + 2q^{49} f(q^6, q^{34})f(q^{114}, q^{646})$$
$$+ 2q^{64} f(q^4, q^{36})f(q^{76}, q^{684}) + 2q^{81} f(q^2, q^{38})f(q^{38}, q^{722}) + q^{100} f(1, q^{40})f(1, q^{760}),$$

after several applications of (2.5). Upon replacing q by $-q$ in (5.16.4), we conclude that

(5.16.5)
$$\frac{1}{4q} \left(\varphi(q)\varphi(q^{19}) - \varphi(-q)\varphi(-q^{19}) \right)$$
$$= f(q^{18}, q^{22})f(q^{342}, q^{418}) + q^8 f(q^{14}, q^{26})f(q^{266}, q^{494}) + q^{24} f(q^{10}, q^{30})f(q^{190}, q^{570})$$
$$+ q^{48} f(q^6, q^{34})f(q^{114}, q^{646}) + q^{80} f(q^2, q^{38})f(q^{38}, q^{722}).$$

Next, we employ (4.14) with the two sets of parameters $\alpha_1 = 1$, $\beta_1 = 19$, $m_1 = 1$, $p_1 = 2$, $\lambda_1 = 10$ and $\alpha_2 = 1$, $\beta_2 = 19$, $m_2 = 9$, $p_2 = 10$, $\lambda_2 = 10$. We find that the conditions in (4.9) are satisfied. Hence, using (4.18) and (4.14), we find that

$$q^{5/2}\psi(-q)\psi(-q^{19}) = q^{5/2}f(-q^{19},-q)f(-q^{209},-q^{171})$$
$$+ q^{45/2}f(-q^{37},-q^{-17})f(-q^{247},-q^{133}) + q^{125/2}f(-q^{55},-q^{-35})f(-q^{285},-q^{95})$$
$$+ q^{245/2}f(-q^{73},-q^{-53})f(-q^{323},-q^{57}) + q^{405/2}f(-q^{91},-q^{-71})f(-q^{361},-q^{19}).$$

After several applications of (2.5), we obtain the identity

$$\psi(-q)\psi(-q^{19}) = f(-q,-q^{19})f(-q^{171},-q^{209}) - q^3 f(-q^3,-q^{17})f(-q^{133},-q^{247})$$
$$+ q^{10}f(-q^5,-q^{15})f(-q^{95},-q^{285}) - q^{21}f(-q^7,-q^{13})f(-q^{57},-q^{323})$$
(5.16.6) $$+ q^{36}f(-q^9,-q^{11})f(-q^{19},-q^{361}).$$

By using (5.16.5) with q replaced by $q^{1/2}$ and (5.16.6) with q replaced by $-q$, we conclude that

$$\frac{1}{4\sqrt{q}}\left(\varphi(q^{1/2})\varphi(q^{19/2}) - \varphi(-q^{1/2})\varphi(-q^{19/2})\right) - q^2\psi(q)\psi(q^{19})$$
$$= f(q^9,q^{11})f(q^{171},q^{209}) + q^4 f(q^7,q^{13})f(q^{133},q^{247}) + q^{24}f(q^3,q^{17})f(q^{57},q^{323})$$
$$+ q^{40}f(q,q^{19})f(q^{19},q^{361}) - q^2 f(q,q^{19})f(q^{171},q^{209}) - q^5 f(q^3,q^{17})f(q^{133},q^{247})$$
$$- q^{23}f(q^7,q^{13})f(q^{57},q^{323}) - q^{38}f(q^9,q^{11})f(q^{19},q^{361})$$
$$= \left(f(q^9,q^{11}) - q^2 f(q,q^{19})\right)\left(f(q^{171},q^{209}) - q^{38}f(q^{19},q^{361})\right)$$
$$+ q^4\left(f(q^7,q^{13}) - qf(q^3,q^{17})\right)\left(f(q^{133},q^{247}) - q^{19}f(q^{57},q^{323})\right)$$
$$= f(-q^2,-q^3)f(-q^{38},-q^{57}) + q^4 f(-q,-q^4)f(-q^{19},-q^{76}),$$

where in the last step we used (5.22.4) and (5.22.5) with q replaced by $-q$ and $-q^{19}$, respectively. This completes the proof of Entry 3.17.

5.17. Proof of Entry 3.18

The following proof of Entry 3.18 is due to Bressoud [14].

By (2.11), (2.7), (2.8), and (2.14), it is easy to see that (3.21) is equivalent to the identity

$$f(-q,-q^4)f(-q^{62},-q^{93}) - q^6 f(-q^2,-q^3)f(-q^{31},-q^{124})$$
(5.17.1) $$= \frac{1}{2q}\varphi(-q^2)\varphi(-q^{62}) - \frac{1}{2q}\varphi(-q)\varphi(-q^{31}) + q^3\psi(-q)\psi(-q^{31}).$$

By (4.16), and (4.13), with the set of parameters, $\alpha = 1/2$, $\beta = 31/2$, $m = 3$, $p = 5$, and $\lambda = 4$, we find that

$$qf(-q,-q^4)f(-q^{62},-q^{93}) - q^7 f(-q^2,-q^3)f(-q^{31},-q^{124})$$
(5.17.2) $$= \sum_{k=1}^{2} F(\tfrac{1}{2},\tfrac{31}{2},3,5,4,k) = \frac{1}{2}\sum_{u,t=-\infty}^{\infty}(-1)^t q^I =: \frac{1}{2}R,$$

where, by (4.7), I is given by

(5.17.3) $$I = 4\left\{u + \frac{1}{2} + \frac{3t}{8}\right\}^2 + \frac{31}{16}t^2 = (2u+1)^2 + \frac{3}{2}(2u+1)t + \frac{5}{2}t^2.$$

Therefore, by (5.17.1)–(5.17.3), it suffices to prove that

$$R = \sum_{u,t=-\infty}^{\infty} (-1)^t q^{(2u+1)^2 + \frac{3}{2}(2u+1)t + \frac{5}{2}t^2}$$

(5.17.4)
$$= \varphi(-q^2)\varphi(-q^{62}) - \varphi(-q)\varphi(-q^{31}) + 2q^4\psi(-q)\psi(-q^{31}).$$

We establish (5.17.4) by a series of changes of the indices of summation. To that end

$$R = \sum_{u,t=-\infty}^{\infty} (-1)^t q^{(2u+1)^2 + \frac{3}{2}(2u+1)t + \frac{5}{2}t^2}$$

$$= \sum_{j=0}^{1} \sum_{u,r=-\infty}^{\infty} (-1)^{2r+j} q^{(2u+1)^2 + \frac{3}{2}(2u+1)(2r+j) + \frac{5}{2}(2r+j)^2}$$

$$= \sum_{u,r=-\infty}^{\infty} q^{4u^2 + 10r^2 + 6ru + 4u + 3r + 1} - \sum_{u,r=-\infty}^{\infty} q^{4u^2 + 10r^2 + 6ru + 7u + 13r + 5}$$

$$= \sum_{s,r=-\infty}^{\infty} q^{4(s-r)^2 + 10r^2 + 6r(s-r) + 4(s-r) + 3r + 1}$$

$$- \sum_{s,r=-\infty}^{\infty} q^{4(s-r-1)^2 + 10r^2 + 6r(s-r-1) + 7(s-r-1) + 13r + 5}$$

$$= \sum_{s,r=-\infty}^{\infty} q^{(2s+1)^2 - (2s+1)r + 8r^2} - \sum_{s,r=-\infty}^{\infty} q^{4s^2 - s(2r+1) + 2(2r+1)^2}$$

(5.17.5)
$$= \sum_{\substack{s,r=-\infty \\ s \text{ odd}}}^{\infty} q^{s^2 - sr + 8r^2} - \sum_{\substack{s,r=-\infty \\ r \text{ odd}}}^{\infty} q^{4s^2 - sr + 2r^2}.$$

But, trivially,

(5.17.6)
$$\sum_{\substack{s,r=-\infty \\ r \text{ even}}}^{\infty} q^{4s^2 - sr + 2r^2} - \sum_{\substack{s,r=-\infty \\ s \text{ even}}}^{\infty} q^{s^2 - sr + 8r^2} = 0.$$

Therefore, returning to (5.17.5), we find that

$$R = \sum_{\substack{s,r=-\infty \\ s \text{ odd}}}^{\infty} q^{s^2 - sr + 8r^2} - \sum_{\substack{s,r=-\infty \\ r \text{ odd}}}^{\infty} q^{4s^2 - sr + 2r^2}$$

$$+ \sum_{\substack{s,r=-\infty \\ r \text{ even}}}^{\infty} q^{4s^2 - sr + 2r^2} - \sum_{\substack{s,r=-\infty \\ s \text{ even}}}^{\infty} q^{s^2 - sr + 8r^2}$$

(5.17.7)
$$= \sum_{s,r=-\infty}^{\infty} (-1)^r q^{4s^2 - sr + 2r^2} - \sum_{s,r=-\infty}^{\infty} (-1)^s q^{s^2 - sr + 8r^2} =: R_1 - R_2.$$

Next, we evaluate R_1 and R_2 separately. First,

$$R_1 = \sum_{s,r=-\infty}^{\infty} (-1)^r q^{4s^2-sr+2r^2} = \sum_{j=0}^{3} \sum_{t,r=-\infty}^{\infty} (-1)^r q^{4(4t+j)^2-(4t+j)r+2r^2}$$

$$= \sum_{j=0}^{3} \sum_{t,r=-\infty}^{\infty} (-1)^r q^{64t^2+32tj+4j^2-4rt-jr+2r^2}$$

$$= \sum_{j=0}^{3} \sum_{t,r=-\infty}^{\infty} (-1)^r q^{62(t+\frac{j}{4})^2 + 2(t-r+\frac{j}{4})^2}$$

$$= \sum_{j=0}^{3} \sum_{t,s=-\infty}^{\infty} (-1)^{t-s} q^{62(t+\frac{j}{4})^2 + 2(s+\frac{j}{4})^2}$$

(5.17.8)
$$= \sum_{j=0}^{3} \left\{ \sum_{t=-\infty}^{\infty} (-1)^t q^{62(t+\frac{j}{4})^2} \sum_{s=-\infty}^{\infty} (-1)^s q^{2(s+\frac{j}{4})^2} \right\}.$$

Observe that

$$\sum_{s=-\infty}^{\infty} (-1)^s q^{(s+\frac{1}{4})^2} = \sum_{s=-\infty}^{\infty} (-1)^s q^{(s+1-\frac{3}{4})^2} = \sum_{s=-\infty}^{\infty} (-1)^{s-1} q^{(s-\frac{3}{4})^2}$$

(5.17.9)
$$= -\sum_{s=-\infty}^{\infty} (-1)^{-s} q^{(-s-\frac{3}{4})^2} = -\sum_{s=-\infty}^{\infty} (-1)^s q^{(s+\frac{3}{4})^2}.$$

Thus, in (5.17.8), the contributions from the terms when $j=1$ and $j=3$ are the same. By (2.4), the contribution from the term $j=2$ is 0. Therefore, by (2.1), (2.7) and (2.8), we conclude that

(5.17.10)
$$R_1 = \sum_{t=-\infty}^{\infty} (-1)^t q^{62t^2} \sum_{s=-\infty}^{\infty} (-1)^s q^{2s^2} + 2q^4 \sum_{t=-\infty}^{\infty} (-1)^t q^{62t^2+31t} \sum_{s=-\infty}^{\infty} (-1)^s q^{2s^2+s}$$
$$= \varphi(-q^2)\varphi(-q^{62}) + 2q^4 f(-q,-q^3)f(-q^{31},-q^{93})$$
$$= \varphi(-q^2)\varphi(-q^{62}) + 2q^4 \psi(-q)\psi(-q^{31}).$$

Similarly,

$$R_2 = \sum_{s,r=-\infty}^{\infty} (-1)^s q^{s^2-sr+8r^2} = \sum_{j=0}^{1} \sum_{s,t=-\infty}^{\infty} (-1)^s q^{s^2-s(2t+j)+8(2t+j)^2}$$

$$= \sum_{j=0}^{1} \sum_{s,t=-\infty}^{\infty} (-1)^s q^{s^2-sj-2st+32t^2+32tj+8j^2}$$

$$= \sum_{j=0}^{1} \sum_{s,t=-\infty}^{\infty} (-1)^s q^{31(t+\frac{j}{2})^2 + (t-s+\frac{j}{2})^2}$$

$$= \sum_{j=0}^{1} \sum_{t,r=-\infty}^{\infty} (-1)^{t-r} q^{31(t+\frac{j}{2})^2 + (r+\frac{j}{2})^2}$$

$$= \sum_{j=0}^{1} \left\{ \sum_{t=-\infty}^{\infty} (-1)^t q^{31(t+\frac{j}{2})^2} \sum_{r=-\infty}^{\infty} (-1)^r q^{(r+\frac{j}{2})^2} \right\}$$

(5.17.11)
$$= \sum_{t=-\infty}^{\infty} (-1)^t q^{31 t^2} \sum_{r=-\infty}^{\infty} (-1)^r q^{r^2} = \varphi(-q)\varphi(-q^{31}),$$

where we used (2.4) again. By (5.17.7), (5.17.10), and (5.17.11), we conclude that
$$R = \varphi(-q^2)\varphi(-q^{62}) + 2q^4 \psi(-q)\psi(-q^{31}) - \varphi(-q)\varphi(-q^{31}),$$
which is (5.17.4). Hence, the proof of Entry 3.18 is complete.

5.18. Proof of Entry 3.19

In the proof below, we actually provide two variations. Like Bressoud [14], we begin with an application of Rogers's ideas, but then the proofs diverge.

Proof. By (2.11), the first part of Entry 3.19 can be put in the form

$$f(-q^2, -q^3)f(-q^{78}, -q^{117}) + q^8 f(-q, -q^4)f(-q^{39}, -q^{156})$$
(5.18.1)
$$= f(-q^{26}, -q^{39})f(-q^3, -q^{12}) - q^2 f(-q^6, -q^9)f(-q^{13}, -q^{52}).$$

We apply Rogers's method first with $\alpha_1 = \frac{1}{2}$, $\beta_1 = \frac{39}{2}$, $p_1 = 5$, $m_1 = 1$, and $\lambda_1 = 4$, and secondly with $\alpha_2 = \frac{3}{2}$, $\beta_2 = \frac{13}{2}$, $p_2 = 5$, $m_2 = 3$, and $\lambda_2 = 4$. Then both sets of parameters satisfy (4.9). By (4.15) and (4.16), respectively, we find that

(5.18.2)
$$\sum_{k=1}^{2} F\left(\tfrac{1}{2}, \tfrac{39}{2}, 1, 5, 4, k\right) = qf(-q^2, -q^3)f(-q^{78}, -q^{117}) + q^9 f(-q, -q^4)f(-q^{39}, -q^{156})$$

and

(5.18.3)
$$\sum_{k=1}^{2} F\left(\tfrac{3}{2}, \tfrac{13}{2}, 3, 5, 4, k\right) = qf(-q^3, -q^{12})f(-q^{26}, -q^{39}) - q^3 f(-q^6, -q^9)f(-q^{13}, -q^{52}).$$

Combining (5.18.2) and (5.18.3) together, we deduce (5.18.1) to complete the proof.

Next, we prove (3.23). Let us define, by (2.11),
(5.18.4)
$$g(q) := f(-q)G(q) = f(-q^2, -q^3) \quad \text{and} \quad h(q) := f(-q)H(q) = f(-q, -q^4).$$

Therefore, by (5.18.1), we can define $N(q)$ by

(5.18.5) $\quad N(q) := g(q)g(q^{39}) + q^8 h(q)h(q^{39}) = g(q^{13})h(q^3) - q^2 g(q^3)h(q^{13}).$

Let us also define
(5.18.6) $\qquad M(q) := g(q^2)g(q^{13}) + q^3 h(q^2)h(q^{13}),$
(5.18.7) $\qquad L(q) := g(q^{26})h(q) - q^5 g(q)h(q^{26}).$

LEMMA 5.7.

$$g(q) = \frac{1}{\varphi(-q^9)} \left\{ -q^2 \varphi(-q)h(q^9) + \varphi(-q^3)\chi(-q^3)g(q^6) \right\},$$

(5.18.8)
$$h(q) = \frac{1}{\varphi(-q^9)} \left\{ \varphi(-q)g(q^9) + q\varphi(-q^3)\chi(-q^3)h(q^6) \right\}.$$

Proof. To prove (5.18.8) we employ (2.19) with $a = -q^2$, $b = -q^3$, and $n = 3$, we find that

$$f(-q^2, -q^3) = f(-q^{21}, -q^{24}) - q^2 f(-q^{36}, -q^9) + q^9 f(-q^{51}, -q^{-6})$$
(5.18.9)
$$= f(-q^{21}, -q^{24}) - q^2 f(-q^9, -q^{36}) - q^3 f(-q^6, -q^{39}),$$

where in the last step (2.5) is used. Similarly, with the choice of parameters $a = -q$, $b = -q^4$, and $n = 3$ one obtains

(5.18.10) $\quad f(-q, -q^4) = f(-q^{18}, -q^{27}) - qf(-q^{12}, -q^{33}) - q^4 f(-q^3, -q^{42}).$

Therefore, in the notation of (5.18.4), we obtain

(5.18.11) $\quad g(q) = A(q^3) - q^2 h(q^9) \text{ and } h(q) = g(q^9) - qB(q^3),$

where
(5.18.12)
$A(q) = f(-q^7, -q^8) - qf(-q^2, -q^{13}) \text{ and } B(q) = f(-q^4, -q^{11}) + qf(-q, -q^{14}).$

Next, we use Entries 3.6–3.8. By (2.14), we can rewrite them in their equivalent forms

(5.18.13) $\quad g(q)g(q^9) + q^2 h(q)h(q^9) = f^2(-q^3),$

(5.18.14) $\quad g(q^2)g(q^3) + qh(q^2)h(q^3) = \psi(q)\varphi(-q^3),$

and

(5.18.15) $\quad h(q)g(q^6) - qg(q)h(q^6) = \psi(q^3)\varphi(-q).$

Starting from (5.18.13), we have, by (5.18.11),

$$f^2(-q^3) = g(q)g(q^9) + q^2 h(q)h(q^9)$$
$$= (A(q^3) - q^2 h(q^9)) g(q^9) + q^2 (g(q^9) - qB(q^3)) h(q^9)$$
(5.18.16)
$$= A(q^3)g(q^9) - q^3 B(q^3)h(q^9).$$

Similarly, starting from (5.18.15) and using (5.18.11) and (5.18.14) with q replaced by q^3, we deduce that

$$\psi(q^3)\varphi(-q)$$
$$= h(q)g(q^6) - qg(q)h(q^6)$$
$$= g(q^6)g(q^9) + q^3 h(q^6)h(q^9) - q \left(A(q^3)h(q^6) + B(q^3)g(q^6)\right)$$
(5.18.17)
$$= \psi(q^3)\varphi(-q^9) - q \left(A(q^3)h(q^6) + B(q^3)g(q^6)\right).$$

Solving for $A(q)$ from the last two equations, we find that

$$\left(g(q^6)g(q^9) + q^3 h(q^6)h(q^9)\right) A(q^3)$$
$$= f^2(-q^3)g(q^6) + q^2 \psi(q^3) \left(\varphi(-q^9) - \varphi(-q)\right) h(q^9).$$

Using (5.18.14) again with q replaced by q^3, we conclude that

(5.18.18) $\quad \varphi(-q^9)A(q^3) = \dfrac{f^2(-q^3)}{\psi(q^3)} g(q^6) + q^2 \left(\varphi(-q^9) - \varphi(-q)\right) h(q^9).$

Substituting this value of $A(q)$ in (5.18.11) yields the first identity of (5.18.8) after observing that

$$\frac{f^2(-q^3)}{\psi(q^3)} = \varphi(-q^3)\chi(-q^3).$$

Similarly one solves for $B(q)$ and obtains the analogous identity for $h(q)$. $\square$

Using (5.18.8) in the first equation of (5.18.5), we find that

$$N(q) = \frac{1}{\varphi(-q^9)} \left\{-q^2\varphi(-q)h(q^9) + \varphi(-q^3)\chi(-q^3)g(q^6)\right\} g(q^{39})$$
$$+ \frac{q^8}{\varphi(-q^9)} \left\{\varphi(-q)g(q^9) + q\varphi(-q^3)\chi(-q^3)h(q^6)\right\} h(q^{39})$$
$$= -q^2 \frac{\varphi(-q)}{\varphi(-q^9)} \left\{g(q^{39})h(q^9) - q^6 g(q^9)h(q^{39})\right\}$$
$$+ \frac{\varphi(-q^3)\chi(-q^3)}{\varphi(-q^9)} \left\{g(q^6)g(q^{39}) + q^9 h(q^6)h(q^{39})\right\}$$
(5.18.19) $\qquad = -q^2 \frac{\varphi(-q)}{\varphi(-q^9)} N(q^3) + \frac{\varphi(-q^3)\chi(-q^3)}{\varphi(-q^9)} M(q^3).$

Employing (5.18.8) again, this time with q replaced by q^{13} in the second equation of (5.18.5), we find that

$$N(q) = \frac{1}{\varphi(-q^{117})} \left\{-q^{26}\varphi(-q^{13})h(q^{117}) + \varphi(-q^{39})\chi(-q^{39})g(q^{78})\right\} h(q^3)$$
$$- \frac{q^2}{\varphi(-q^{117})} \left\{\varphi(-q^{13})g(q^{117}) + q^{13}\varphi(-q^{39})\chi(-q^{39})h(q^{78})\right\} g(q^3)$$
$$= -q^2 \frac{\varphi(-q^{13})}{\varphi(-q^{117})} \left\{g(q^3)g(q^{117}) + q^{24} h(q^3)h(q^{117})\right\}$$
$$+ \frac{\varphi(-q^{39})\chi(-q^{39})}{\varphi(-q^{117})} \left\{h(q^3)g(q^{78}) - q^{15} g(q^3)h(q^{78})\right\}$$
(5.18.20) $\qquad = -q^2 \frac{\varphi(-q^{13})}{\varphi(-q^{117})} N(q^3) + \frac{\varphi(-q^{39})\chi(-q^{39})}{\varphi(-q^{117})} L(q^3).$

From (3.18), (2.11), and (2.14), we deduce that

(5.18.21) $\qquad \dfrac{L(q)}{M(q)} = \dfrac{f(-q)f(-q^{26})}{f(-q^2)f(-q^{13})} = \dfrac{\chi(-q)}{\chi(-q^{13})}.$

Thus, by (5.18.19)–(5.18.21), we conclude that

(5.18.22)
$$q^2 \left\{\frac{\varphi(-q^{13})}{\varphi(-q^{117})} - \frac{\varphi(-q)}{\varphi(-q^9)}\right\} N(q^3) = \left\{\frac{\varphi(-q^{39})\chi(-q^3)}{\varphi(-q^{117})} - \frac{\varphi(-q^3)\chi(-q^3)}{\varphi(-q^9)}\right\} M(q^3).$$

By (5.28.23), with q replaced by q^{13} and q, respectively, we find that

$$\varphi(-q^9)\varphi(-q^{13}) - \varphi(-q)\varphi(-q^{117})$$
$$= \varphi(-q^9)\left\{\varphi(-q^{117}) - 2q^{13}f(-q^{39}, -q^{195})\right\}$$
$$- \left\{\varphi(-q^9) - 2qf(-q^3, -q^{15})\right\}\varphi(-q^{117})$$
(5.18.23) $\qquad = 2q\left\{f(-q^3, -q^{15})\varphi(-q^{117}) - q^{12}\varphi(-q^9)f(-q^{39}, -q^{195})\right\}.$

Using (5.18.23) in (5.18.22) and replacing by q^3 by q, we arrive at

$$2q\left\{f(-q, -q^5)\varphi(-q^{39}) - q^4\varphi(-q^3)f(-q^{13}, -q^{65})\right\} N(q)$$
(5.18.24) $\qquad = \chi(-q)\left\{\varphi(-q^3)\varphi(-q^{13}) - \varphi(-q)\varphi(-q^{39})\right\} M(q).$

Comparing (5.18.24) to (3.23), we see that it suffices to prove that

$$M(q) = \frac{1}{\chi(-q)}\{f(-q,-q^5)\varphi(-q^{39}) - q^4\varphi(-q^3)f(-q^{13},-q^{65})\}$$
(5.18.25)
$$= \psi(q^3)\varphi(-q^{39}) - q^4 f(q,q^2)f(-q^{13},-q^{65}),$$

where in the last step we used (5.7.6) and (5.7.8). To verify (5.18.25), we employ Theorem 4.1 with the parameters $a = b = q^{39}$, $c = 1$, $d = q^3$, $\epsilon_1 = 1$, $\epsilon_2 = 0$, $\alpha = 2$, $\beta = 1$, and $m = 15$, to find that

(5.18.26)
$$f(1,q^3)\varphi(-q^{39}) = 2\,f(-q^{42},-q^{48})f(-q^{273},-q^{312})$$
$$+ 2\,q^3 f(-q^{36},-q^{54})f(-q^{234},-q^{351}) + 2\,q^9 f(-q^{30},-q^{60})f(-q^{195},-q^{390})$$
$$+ 2\,q^{18} f(-q^{24},-q^{66})f(-q^{156},-q^{429}) + 2\,q^{30} f(-q^{18},-q^{72})f(-q^{117},-q^{468})$$
$$+ 2\,q^{45} f(-q^{12},-q^{78})f(-q^{78},-q^{507}) + 2\,q^{63} f(-q^6,-q^{84})f(-q^{39},-q^{546}).$$

Employing Theorem 4.1 again, this time with the parameters $a = q^{13}$, $b = q^{65}$, $c = q$, $d = q^2$, $\epsilon_1 = 1$, $\epsilon_2 = 0$, $\alpha = 13$, $\beta = 1$, and $m = 15$, we find that

(5.18.27)
$$f(-q^{13},-q^{65})f(q,q^2) = f(-q^{273},-q^{312})f(-q^{18},-q^{72})$$
$$+ qf(-q^{234},-q^{351})f(-q^{24},-q^{66}) + q^5 f(-q^{195},-q^{390})f(-q^{30},-q^{60})$$
$$+ q^{12} f(-q^{156},-q^{429})f(-q^{36},-q^{54}) + q^{22} f(-q^{117},-q^{468})f(-q^{42},-q^{48})$$
$$+ q^{35} f(-q^{78},-q^{507})f(-q^{42},-q^{48}) + q^{51} f(-q^{39},-q^{546})f(-q^{36},-q^{54})$$
$$- q^{53} f(-q^{39},-q^{546})f(-q^{24},-q^{66}) - q^{39} f(-q^{78},-q^{507})f(-q^{18},-q^{72})$$
$$- q^{28} f(-q^{117},-q^{468})f(-q^{12},-q^{78}) - q^{20} f(-q^{156},-q^{429})f(-q^6,-q^{84})$$
$$+ q^7 f(-q^{234},-q^{351})f(-q^6,-q^{84}) + q^2 f(-q^{273},-q^{312})f(-q^{12},-q^{78}).$$

Now by (2.3), (5.18.26), and (5.18.27), we conclude that

(5.18.28)
$$\psi(q^3)\varphi(-q^{39}) - q^4 f(q,q^2)f(-q^{13},-q^{65})$$
$$= \{f(-q^{42},-q^{48}) - q^4 f(-q^{18},-q^{72}) - q^6 f(-q^{12},-q^{78})\}\{f(-q^{273},-q^{312})$$
$$- q^{26} f(-q^{117},-q^{468}) - q^{39} f(-q^{78},-q^{507})\}$$
$$+ q^3 \{f(-q^{36},-q^{54}) - q^2 f(-q^{24},-q^{66}) - q^8 f(-q^6,-q^{84})\}\{f(-q^{234},-q^{351})$$
$$- q^{13} f(-q^{156},-q^{429}) - q^{52} f(-q^{39},-q^{546})\}.$$

But, from (2.19) with $n = 3$, we know that

(5.18.29)
$$g(q) = f(-q^2,-q^3) = f(-q^{21},-q^{24}) - q^2 f(-q^9,-q^{36}) - q^3 f(-q^6,-q^{39}),$$
(5.18.30)
$$h(q) = f(-q,-q^4) = f(-q^{18},-q^{27}) - qf(-q^{12},-q^{33}) - q^4 f(-q^3,-q^{42}),$$

where we used (2.5). Replacing q by q^2 and q^{13} in each of (5.18.29) and (5.18.30), we see that (5.18.25) holds, since the right-hand side of (5.18.28) is exactly

$$g(q^2)g(q^{13}) + q^3 h(q^2)h(q^{13}) = M(q).$$

Hence, the proof of Entry 3.19 is complete. □

Next, we sketch a different prove for Entry 3.19 which, by (5.18.21), is equivalent to showing that

$$\frac{L(q)}{M(q)} = \frac{\chi(-q)}{\chi(-q^{13})}. \tag{5.18.31}$$

Therefore, by (5.18.25), one needs to prove that

$$\begin{aligned}L(q) &= \frac{\chi(-q)}{\chi(-q^{13})}\left\{\psi(q^3)\varphi(-q^{39}) - q^4 f(q,q^2)f(-q^{13},-q^{65})\right\} \\ &= f(-q,-q^5)f(q^{13},q^{26}) - q^4\varphi(-q^3)\psi(q^{39}),\end{aligned} \tag{5.18.32}$$

where in the last step we used (5.7.6) and (5.7.8). The equality (5.18.32) is proved in the same way that we proved (5.18.25), and so we omit the details.

5.19. Proofs of Entry 3.20

Proof. Using (4.23) and (4.24) in (3.3), we arrive at

$$\begin{aligned}\chi^2(q) &= G(q)G(q^4) + qH(q)H(q^4) \\ &= \frac{f(-q^8)}{f(-q^2)}\left\{G(q^4)\left(G(q^{16}) + qH(-q^4)\right) + qH(q^4)\left(q^3 H(q^{16}) + G(-q^4)\right)\right\} \\ &= \frac{f(-q^8)}{f(-q^2)}\left\{G(q^4)G(q^{16}) + q^4 H(q^4)H(q^{16})\right. \\ &\quad \left. + q\left(H(-q^4)G(q^4) + H(q^4)G(-q^4)\right)\right\}.\end{aligned} \tag{5.19.1}$$

Separating the even and odd indexed terms, we easily show that

$$\varphi(q) = \varphi(q^4) + 2q\psi(q^8). \tag{5.19.2}$$

Using (2.10), (2.7), and (5.19.2), we conclude from (5.19.1) that

$$\begin{aligned}&G(q^4)G(q^{16}) + q^4 H(q^4)H(q^{16}) + q\left(H(-q^4)G(q^4) + H(q^4)G(-q^4)\right) \\ &= \frac{\chi^2(q)f(-q^2)}{f(-q^8)} = \frac{\varphi(q)}{f(-q^8)} = \frac{\varphi(q^4) + 2q\psi(q^8)}{f(-q^8)}.\end{aligned} \tag{5.19.3}$$

Equating odd parts on both sides of (5.19.3), we deduce that

$$H(-q^4)G(q^4) + H(q^4)G(-q^4) = 2\frac{\psi(q^8)}{f(-q^8)}, \tag{5.19.4}$$

which is Entry 3.20 with q replaced by q^4. □

5.20. Proof of Entry 3.21

We shall see that Watson's proof [34] of Entry 3.21 follows by combining Entries 3.1 and 3.2 with some elementary identities for theta functions.

From Entries 3.2 and 3.3, we easily deduce that

$$G(q) = \frac{\varphi(q) + \varphi(q^5)}{2G(q^4)f(-q^2)} \quad \text{and} \quad qH(q) = \frac{\varphi(q) - \varphi(q^5)}{2H(q^4)f(-q^2)}. \tag{5.20.1}$$

Applying each of the equalities in (5.20.1) twice, but with q replaced by $-q$ in two instances, using (2.16), using Lemma 5.2, and invoking (2.12), we find that

$$qG(q)H(-q) - qG(-q)H(q) = \frac{\varphi(q^5)\varphi(-q^5) - \varphi(q)\varphi(-q)}{2G(q^4)H(q^4)f^2(-q^2)}$$

$$= \frac{\varphi^2(-q^{10}) - \varphi^2(-q^2)}{2G(q^4)H(q^4)f^2(-q^2)}$$

$$= \frac{2q^2\chi(-q^2)f^2(-q^{20})}{G(q^4)H(q^4)\chi(-q^{10})f^2(-q^2)}$$

$$= \frac{2q^2 f(-q^{20})}{\chi(-q^{10})f(-q^2)} \cdot \frac{f(-q^4)\chi(-q^2)}{f(-q^2)}$$

(5.20.2) $$= \frac{2q^2 \psi(q^{10})}{f(-q^2)},$$

where we applied the elementary identities

(5.20.3) $$\psi(q)\chi(-q) = f(-q^2) = \frac{f(-q)}{\chi(-q)},$$

with q replaced by q^{10} and q^2, respectively. The identities in (5.20.3) both follow from (2.14). The truth of Entry 3.21 is readily apparent from (5.20.2).

5.21. Proof of Entry 3.22

First Proof of Entry 3.22. Using (4.23) and (4.24) in Entry 3.5, we find that

$$\chi(q^2) = G(q^{16})H(q) - q^3 G(q)H(q^{16})$$

$$= \left\{ \frac{f(-q^2)}{f(-q^8)} G(q) - qH(-q^4) \right\} H(q) - \left\{ \frac{f(-q^2)}{f(-q^8)} H(q) - G(-q^4) \right\} G(q)$$

$$= G(q)G(-q^4) - qH(q)H(-q^4),$$

which is Entry 3.22 with q replaced by $-q$. $\square$

Second Proof of Entry 3.22. Consider the system of three equations,

(5.21.1) $$G(-q)G(-q^4) + qH(-q)H(-q^4) =: T(q),$$

(5.21.2) $$H(q^4)G(-q^4) + G(q^4)H(-q^4) = \frac{2\psi(q^8)}{f(-q^8)},$$

(5.21.3) $$-H(q^4)G(-q^4) + G(q^4)H(-q^4) = \frac{2q^4\psi(q^{40})}{f(-q^8)}.$$

Note that (5.21.1) merely gives the definition of $T(q)$, and that our goal is to show that $T(q) = \chi(q^2)$. The equality (5.21.2) is (3.24) with q replaced by q^4, and (5.21.3) is (3.25) with q replaced by q^4. We regard (5.21.1)–(5.21.3) as a system of three equations in the "variables" $G(-q^4)$, $H(-q^4)$, and -1. Thus, we have

(5.21.4) $$\begin{vmatrix} G(-q) & qH(-q) & T(q) \\ H(q^4) & G(q^4) & \dfrac{2\psi(q^8)}{f(-q^8)} \\ -H(q^4) & G(q^4) & \dfrac{2q^4\psi(q^{40})}{f(-q^8)} \end{vmatrix} = 0.$$

Expanding (5.21.4) by the last column, we find that

$$2T(q)G(q^4)H(q^4) - \frac{2\psi(q^8)}{f(-q^8)}\{G(-q)G(q^4) + qH(-q)H(q^4)\}$$
(5.21.5)
$$+ \frac{2q^4\psi(q^{40})}{f(-q^8)}\{G(-q)G(q^4) - qH(-q)H(q^4)\} = 0.$$

Using (2.12), (3.4) with q replaced by $-q$, and (3.3) with q replaced by $-q$, we rewrite (5.21.5) in the form

(5.21.6)
$$\frac{T(q)f(-q^{20})}{f(-q^4)} - \frac{\psi(q^8)\varphi(-q^5)}{f(-q^8)f(-q^2)} + \frac{q^4\psi(q^{40})\varphi(-q)}{f(-q^8)f(-q^2)} = 0,$$

or, upon rearrangement,

(5.21.7) $\quad T(q) = \dfrac{f(-q^4)}{f(-q^2)f(-q^8)f(-q^{20})}\left\{\varphi(-q^5)\psi(q^8) - q^4\varphi(-q)\psi(q^{40})\right\}.$

By (2.15), (5.21.7) can be rewritten as

(5.21.8) $\quad T(q) = \dfrac{\chi(-q^2)\chi(q^2)}{f(-q^2)f(-q^{20})}\left\{\varphi(-q^5)\psi(q^8) - q^4\varphi(-q)\psi(q^{40})\right\}.$

From (5.5.12), we find, after simplification, that

$$\varphi(-q^5)\psi(q^8) - q^4\varphi(-q)\psi(q^{40})$$
$$= \sqrt{z_5}(1-\beta)^{1/4}\frac{1}{4q}\sqrt{z_1}\left\{1 - (1-\alpha)^{1/4}\right\}$$
$$- q^4\sqrt{z_1}(1-\alpha)^{1/4}\frac{1}{4q^5}\sqrt{z_5}\left\{1 - (1-\beta)^{1/4}\right\}$$
(5.21.9)
$$= \frac{\sqrt{z_1 z_5}}{4q}\left\{(1-\beta)^{1/4} - (1-\alpha)^{1/4}\right\}.$$

Putting (5.5.15) and (5.21.9) in (5.21.8), we arrive at

(5.21.10) $\quad T(q) = \dfrac{\chi(q^2)\left\{(1-\beta)^{1/4} - (1-\alpha)^{1/4}\right\}}{2^{2/3}(\alpha\beta)^{1/6}\left\{(1-\alpha)(1-\beta)\right\}^{1/24}}.$

In comparing (5.21.10) with (3.26), we see that it remains to show that

(5.21.11) $\quad \dfrac{(1-\beta)^{1/4} - (1-\alpha)^{1/4}}{2^{2/3}(\alpha\beta)^{1/6}\{(1-\alpha)(1-\beta)\}^{1/24}} = 1.$

But (5.21.11) is equivalent to (5.5.1), and so the proof is complete. □

5.22. Proof of Entry 3.23

We first remark that we have already given one proof of Entry 3.23 along with one of our proofs of Entry 3.11. We provide a second proof here.

Using (2.11) and (2.14), we see that Entry 3.23 is equivalent to the identity

$$f(-q^4,q^6)f(-q^6,q^9) + qf(q^2,-q^8)f(q^3,-q^{12}) = \frac{\chi(q)\chi(q^6)}{\chi(q^2)\chi(q^3)}f(q^2)f(q^3)$$
(5.22.1)
$$= f(-q^4)f(-q^6)\chi(q)\chi(q^6).$$

Using the product representations of $\chi(q)$ and $f(-q)$ given in (2.9) and (2.10), respectively, together with (2.6), we find that

$$f(-q^4)\chi(q) = (q^4;q^4)_\infty(-q;q^2)_\infty = (q^4;q^4)_\infty(-q;q^4)_\infty(-q^3;q^4)_\infty$$
$$= f(q,q^3) = \psi(q)$$

and

$$f(-q^6)\chi(q^6) = (q^6;q^6)_\infty(-q^6;q^{12})_\infty = (q^6;q^{12})_\infty(q^{12};q^{12})_\infty(-q^6;q^{12})_\infty$$
$$= (q^{12};q^{24})_\infty(q^{12};q^{12})_\infty$$
$$= (q^{12};q^{24})_\infty(q^{12};q^{24})_\infty(q^{24};q^{24})_\infty$$
$$= f(-q^{12},-q^{12}) = \varphi(-q^{12}),$$

by (2.7). It thus suffices to prove that

(5.22.2) $\qquad f(-q^4,q^6)f(-q^6,q^9) + qf(q^2,-q^8)f(q^3,-q^{12}) = \varphi(-q^{12})\psi(q).$

We now apply Theorem 4.1 with the parameters $\epsilon_1 = 1$, $\epsilon_2 = 0$, $a = b = q^{12}$ $c = q$, $d = q^3$, $\alpha = 2$, $\beta = 1$, and $m = 5$. We consequently find that

$$\varphi(-q^{12})\psi(q) = f(-q^{22},-q^{18})f(-q^{33},-q^{27}) + qf(-q^{14},-q^{26})f(-q^{21},-q^{39})$$
$$+ q^6 f(-q^6,-q^{34})f(-q^9,-q^{51}) + q^{15} f(-q^{-2},-q^{42})f(-q^{-3},-q^{63})$$
$$+ q^{28} f(-q^{-10},-q^{50})f(-q^{-15},-q^{75})$$
$$= f(-q^{18},-q^{22})f(-q^{33},-q^{27}) + qf(-q^{14},-q^{26})f(-q^{21},-q^{39})$$
$$+ q^6 f(-q^6,-q^{34})f(-q^9,-q^{51}) + q^{10} f(-q^2,-q^{38})f(-q^3,-q^{57})$$
(5.22.3) $\qquad + q^3 f(-q^{10},-q^{30})f(-q^{15},-q^{45}),$

where we applied (2.5) four times in the last equality. By (2.19), with $a = -q^2, b = q^3$, and $n = 2$, and with $a = q, b = -q^4$, and $n = 2$, respectively,

(5.22.4) $\qquad f(-q^2,q^3) = f(-q^9,-q^{11}) - q^2 f(-q^{19},-q),$
(5.22.5) $\qquad f(q,-q^4) = f(-q^7,-q^{13}) + qf(-q^{17},-q^3).$

Replacing q by q^2 and q^3 in each of (5.22.4) and (5.22.5), respectively, we find that

$$f(-q^4,q^6) = f(-q^{18},-q^{22}) - q^4 f(-q^{38},-q^2),$$
$$f(-q^6,q^9) = f(-q^{27},-q^{33}) - q^6 f(-q^{57},-q^3),$$
$$f(q^2,-q^8) = f(-q^{14},-q^{26}) + q^2 f(-q^{34},-q^6),$$
$$f(q^3,-q^{12}) = f(-q^{21},-q^{39}) + q^3 f(-q^{51},-q^9).$$

Return to (5.22.3) and substitute each of the equalities above to deduce that

$$\varphi(-q^{12})\psi(q) - \{f(-q^4,q^6)f(-q^6,q^9) + qf(q^2,-q^8)f(q^3,-q^{12})\}$$
$$= q^3 f(-q^{10},-q^{30})f(-q^{15},-q^{45}) + q^4 f(-q^2,-q^{38})f(-q^{27},-q^{33})$$
$$- q^3 f(-q^6,-q^{34})f(-q^{21},-q^{39}) - q^4 f(-q^{14},-q^{26})f(-q^9,-q^{51})$$
(5.22.6) $\qquad + q^6 f(-q^{18},-q^{22})f(-q^3,-q^{57}).$

We now use Theorem 4.1 again, but now with the parameters $\epsilon_1 = 1$, $\epsilon_2 = 0$, $a = 1$, $b = q^{24}$, $c = q$, $d = q^3$, $\alpha = 2$, $\beta = 1$, and $m = 5$. Hence, we find that

$$q^3 f(-1, -q^{24}) \psi(q) = q^3 f(-q^{10}, -q^{30}) f(-q^{15}, -q^{45})$$
$$+ q^4 f(-q^2, -q^{38}) f(-q^{27}, -q^{33}) + q^9 f(-q^{-6}, -q^{46}) f(-q^{39}, -q^{21})$$
$$+ q^{18} f(-q^{-14}, -q^{54}) f(-q^{51}, -q^9) + q^{31} f(-q^{-22}, -q^{62}) f(-q^{63}, -q^{-3})$$
$$= q^3 f(-q^{10}, -q^{30}) f(-q^{15}, -q^{45}) + q^4 f(-q^2, -q^{38}) f(-q^{27}, -q^{33})$$
$$- q^3 f(-q^6, -q^{34}) f(-q^{21}, -q^{39}) - q^4 f(-q^{14}, -q^{26}) f(-q^9, -q^{51})$$
(5.22.7) $\quad + q^6 f(-q^{18}, -q^{22}) f(-q^3, -q^{57}),$

after four applications of (2.5). The product on the far left side of (5.22.7) equals 0, by (2.4). Hence, since the right-hand sides of (5.22.7) and (5.22.6) are equal, we complete the proof of (5.22.2), and hence also of Entry 3.23.

5.23. Proof of Entry 3.24

We first remark that we have already given one proof of Entry 3.24 along with one of our proofs of Entry 3.12. We provide a second proof here.

This proof of Entry 3.24 is very similar to that given above for Entry 3.23. Using (2.11) and (2.14), we see that Entry 3.24 is equivalent to the identity

(5.23.1)
$$f(-q^{12}, q^{18}) f(q, -q^4) - q f(q^6, -q^{24}) f(-q^2, q^3) = \frac{\chi(q^2)\chi(q^3)}{\chi(q)\chi(q^6)} f(q) f(q^6)$$
$$= f(-q^2) f(-q^{12}) \chi(q^2) \chi(q^3).$$

Using the product representations of $\chi(q)$ and $f(-q)$ from (2.10) and (2.9), respectively, together with (2.6), we obtain

$$f(-q^2) f(-q^{12}) \chi(q^2) \chi(q^3) = (q^2; q^2)_\infty (q^{12}; q^{12})_\infty (-q^2; q^4)_\infty (-q^3; q^6)_\infty$$
$$= \frac{(q^4; q^8)_\infty}{(q^2; q^4)_\infty} (-q^3; q^{12})_\infty (-q^9; q^{12})_\infty (q^{12}; q^{12})_\infty (q^2; q^4)_\infty (q^4; q^4)_\infty$$
$$= f(q^3, q^9)(q^4; q^8)_\infty (q^4; q^4)_\infty$$
$$= f(q^3, q^9) f(-q^4, -q^4) = \psi(q^3) \varphi(-q^4).$$

It therefore remains to prove that

(5.23.2) $\quad f(-q^{12}, q^{18}) f(q, -q^4) - q f(q^6, -q^{24}) f(-q^2, q^3) = \varphi(-q^4) \psi(q^3).$

We now apply Theorem 4.1 with the parameters $\epsilon_1 = 1$, $\epsilon_2 = 0$, $a = b = q^4$, $c = q^3$, $d = q^9$, $\alpha = 1$, $\beta = 3$, and $m = 5$. Accordingly, we find that

(5.23.3)
$$\varphi(-q^4) \psi(q^3) = f(-q^{13}, -q^7) f(-q^{66}, -q^{54}) + q^3 f(-q, -q^{19}) f(-q^{42}, -q^{78})$$
$$+ q^{18} f(-q^{-11}, -q^{31}) f(-q^{18}, -q^{102}) + q^{45} f(-q^{-23}, -q^{43}) f(-q^{-6}, -q^{126})$$
$$+ q^{84} f(-q^{-35}, -q^{55}) f(-q^{-30}, -q^{150})$$
$$= f(-q^7, -q^{13}) f(-q^{54}, -q^{66}) + q^3 f(-q, -q^{19}) f(-q^{42}, -q^{78})$$
$$- q^7 f(-q^9, -q^{11}) f(-q^{18}, -q^{102}) - q^{13} f(-q^3, -q^{17}) f(-q^6, -q^{114})$$
$$- q^4 f(-q^5, -q^{15}) f(-q^{30}, -q^{90}),$$

where we applied (2.5) five times in the last equality. Recording again (5.22.4) and (5.22.5) as well as their analogues with q replaced by q^6, we find that

$$f(-q^2, q^3) = f(-q^9, -q^{11}) - q^2 f(-q^{19}, -q),$$
$$f(-q^{12}, q^{18}) = f(-q^{54}, -q^{66}) - q^{12} f(-q^{114}, -q^6).$$
$$f(q, -q^4) = f(-q^7, -q^{13}) + q f(-q^{17}, -q^3),$$
$$f(q^6, -q^{24}) = f(-q^{42}, -q^{78}) + q^6 f(-q^{102}, -q^{18}).$$

Using these identities in (5.23.3), we find that, after some elementary algebra,

$$\varphi(-q^4)\psi(q^3) - \{f(-q^{12}, q^{18})f(q, -q^4) - qf(q^6, -q^{24})f(-q^2, q^3)\}$$
$$= qf(-q^9, -q^{11})f(-q^{42}, -q^{78}) - qf(-q^3, -q^{17})f(-q^{54}, -q^{66})$$
$$- q^4 f(-q^5, -q^{15})f(-q^{30}, -q^{90}) + q^{12} f(-q^7, -q^{13})f(-q^6, -q^{114})$$
(5.23.4) $$- q^9 f(-q, -q^{19})f(-q^{18}, -q^{102}).$$

Next, we apply Theorem 4.1 again, but now with the parameters $\epsilon_1 = 1$, $\epsilon_2 = 0$, $a = 1$, $b = q^8$, $c = q^3$, $d = q^9$, $\alpha = 1$, $\beta = 3$, and $m = 5$. Accordingly, we find that

(5.23.5)
$$qf(-1, -q^8)\psi(q^3) = qf(-q^9, -q^{11})f(-q^{78}, -q^{42}) + q^4 f(-q^{-3}, -q^{23})f(-q^{54}, -q^{66})$$
$$+ q^{19} f(-q^{-15}, -q^{35})f(-q^{30}, -q^{90}) + q^{46} f(-q^{-27}, -q^{47})f(-q^6, -q^{114})$$
$$+ q^{85} f(-q^{-39}, -q^{59})f(-q^{138}, -q^{-18})$$
$$= qf(-q^9, -q^{11})f(-q^{42}, -q^{78}) - qf(-q^3, -q^{17})f(-q^{54}, -q^{66})$$
$$- q^4 f(-q^5, -q^{15})f(-q^{30}, -q^{90}) + q^{12} f(-q^7, -q^{13})f(-q^6, -q^{114})$$
$$- q^9 f(-q, -q^{19})f(-q^{18}, -q^{102}),$$

after five applications of (2.5). The right sides of (5.23.4) and (5.23.5) are identical. Thus, the left sides of (5.23.4) and (5.23.5) are identical. Since the left side of (5.23.5) equals 0 by (2.4), we see that (5.23.2) follows immediately. This completes the proof of Entry 3.24.

5.24. Proofs of Entry 3.25

First Proof of Entry 3.25. Using (4.23) and (4.24) in (3.14) with q replaced by q^9, we arrive at

(5.24.1)
$$\frac{\chi(-q)\chi(-q^6)}{\chi(-q^3)\chi(-q^{18})} = G(q^9)H(q^4) - qH(q^9)G(q^4)$$
$$= \frac{f(-q^{72})}{f(-q^{18})} \left\{ H(q^4) \left(G(q^{144}) + q^9 H(-q^{36}) \right) - qG(q^4) \left(q^{27} H(q^{144}) + G(-q^{36}) \right) \right\}$$
$$= \frac{f(-q^{72})}{f(-q^{18})} \left\{ H(q^4)G(q^{144}) - q^{28}G(q^4)H(q^{144}) \right.$$
$$\left. - q \left(G(q^4)G(-q^{36}) - q^8 H(q^4)H(-q^{36}) \right) \right\}.$$

Using (2.14), (2.15), and (5.7.10) with q replaced by $-q$, we deduce from (5.24.1) that

(5.24.2)
$$H(q^4)G(q^{144}) - q^{28}G(q^4)H(q^{144}) - q\left(G(q^4)G(-q^{36}) - q^8 H(q^4)H(-q^{36})\right)$$
$$= \frac{f(-q^{18})\chi(-q)\chi(-q^6)}{\chi(-q^3)\chi(-q^{18})f(-q^{72})} = \chi(-q)\chi(q^3)\chi(-q^{36})$$
$$= \chi(-q^{36})\left\{\frac{\chi(q^{12})}{\chi(-q^8)} - q\frac{\chi(q^4)}{\chi(-q^{24})}\right\}.$$

Equating the even and odd parts in both sides of the equation (5.24.2), we readily obtain Entries 3.14 and 3.25 with q replaced by q^4 and $-q^4$, respectively. □

Second Proof of Entry 3.25. Employing Theorem 4.1 with the set of parameters $a = q^6$, $b = q^{12}$, $c = q$, $d = q^2$, $\alpha = 2$, $\beta = 1$, $m = 5$, $\epsilon_1 = 0$, and $\epsilon_2 = 1$, we find that

$$f(q^6, q^{12})f(-q) = f(q^{13}, q^{17})f(-q^{18}, -q^{27}) - qf(q^7, q^{23})f(-q^{18}, -q^{27})$$
$$+ q^5 f(q, q^{29})f(-q^9, -q^{36}) - q^2 f(q^{11}, q^{19})f(-q^9, -q^{36}),$$

where we used (2.4) and (2.5) twice. Upon the rearrangement of terms and use of (4.26) and (4.27), with q replaced by $-q$, and (2.11), we deduce that

(5.24.3)
$$f(q^6, q^{12})f(-q) = f(-q^{18}, -q^{27})\{f(q^{13}, q^{17}) - qf(q^7, q^{23})\}$$
$$- q^2 f(-q^9, -q^{36})\{f(q^{11}, q^{19}) - q^3 f(q, q^{29})\}$$
$$= f(-q^{18}, -q^{27})G(-q)f(-q^2) - q^2 f(-q^9, -q^{36})H(-q)f(-q^2)$$
$$= f(-q^2)f(-q^9)\left\{G(q^9)G(-q) - q^2 H(q^9)H(-q)\right\}.$$

By (5.7.8) with q replaced by q^6, (2.14), and (2.17) in the form $\chi(q)f(-q) = \varphi(-q^2)$, but with q replaced by q^9, we find that

(5.24.4)
$$\frac{f(-q)f(q^6, q^{12})}{f(-q^2)f(-q^9)} = \chi(-q)\frac{\varphi(-q^{18})}{\chi(-q^6)f(-q^9)} = \frac{\chi(-q)\chi(q^9)}{\chi(-q^6)}.$$

Hence, by (5.24.3) and (5.24.4), the proof of Entry 3.25 is complete. □

Third Proof of Entry 3.25. To prove Entry 3.25, we need the identity [5, p. 349, Entry 2(i)]

(5.24.5) $$\varphi(q)\varphi(q^9) - \varphi^2(q^3) = 2q\varphi(-q^2)\psi(q^9)\chi(q^3).$$

Recall the definitions

$$g(q) = f(-q)G(q) = f(-q^2, -q^3) \quad \text{and} \quad h(q) = f(-q)H(q) = f(-q, -q^4).$$

Using (2.11), (2.7), (2.6), and some elementary product manipulations, we see that Entry 3.25 is equivalent to the identity

(5.24.6) $$g(-q)g(q^9) - q^2 h(-q)h(q^9) = \varphi(-q^2)f(q^6, q^{12}).$$

Replacing q by $-q$ in (5.24.6) gives

(5.24.7) $$g(q)g(-q^9) - q^2 h(q)h(-q^9) = \varphi(-q^2)f(q^6, q^{12}).$$

We prove (5.24.7).

Using (2.11), (2.8), (2.6), and some elementary product manipulations we can express Entry 3.13 as

$$(5.24.8) \qquad g(q^9)h(q^4) - qg(q^4)h(q^9) = \psi(-q)f(q^3, q^{15}).$$

It is also easily verified, using the product expansions from (2.8) and (2.9), that Entry 3.20 is equivalent to the identity

$$(5.24.9) \qquad g(q)h(-q) + g(-q)h(q) = 2\psi(q)\psi(-q).$$

Consider the system of three equations,

$$(5.24.10) \qquad g(q)g(-q^9) - q^2 h(q)h(-q^9) =: T(q),$$

$$(5.24.11) \qquad g(q)g(q^9) + q^2 h(q)h(q^9) = f^2(-q^3),$$

$$(5.24.12) \qquad g(q)g(q^4) + qh(q)h(q^4) = \psi(q)\varphi(-q^2).$$

Equation (5.24.11) above is equation (5.6.1), while equation (5.24.12) is a variation of equation (3.2). We wish to show that $T(q) = \varphi(-q^2)f(q^6, q^{12})$. Regarding this system in the variables $g(q)$, $qh(q)$, and -1, we find that

$$(5.24.13) \qquad \begin{vmatrix} g(-q^9) & -qh(-q^9) & T(q) \\ g(q^9) & qh(q^9) & f^2(-q^3) \\ g(q^4) & h(q^4) & \varphi(-q^2)\psi(q) \end{vmatrix} = 0.$$

Expanding the determinant in (5.24.13) along the last column, we find that

$$(5.24.14)$$
$$T(q)\{g(q^9)h(q^4) - qg(q^4)h(q^9)\} - f^2(-q^3)\{g(-q^9)h(q^4) + qg(q^4)h(-q^9)\}$$
$$+ \varphi(-q^2)\psi(q)\{qg(-q^9)h(q^9) + qg(q^9)h(-q^9)\} = 0.$$

Using (5.24.8), (5.24.8) with q replaced by $-q$, and (5.24.9) with q replaced by q^9 in (5.24.14), we find that

$$T(q)\psi(-q)f(q^3, q^{15}) = f^2(-q^3)\psi(q)f(-q^3, -q^{15}) - 2q\varphi(-q^2)\psi(q)\psi(-q^9)\psi(q^9).$$

It suffices then to prove that

$$(5.24.15) \qquad \begin{aligned} & \varphi(-q^2)f(q^6, q^{12})\psi(-q)f(q^3, q^{15}) \\ & = f^2(-q^3)\psi(q)f(-q^3, -q^{15}) - 2q\varphi(-q^2)\psi(q)\psi(-q^9)\psi(q^9). \end{aligned}$$

By (2.6), (2.7), and (2.8), we find that

$$f(q^2, q^4)f(-q, -q^5)f(-q^2)$$
$$= (-q^2; q^6)_\infty (-q^4; q^6)_\infty (q^6; q^6)_\infty (q; q^6)_\infty (q^5; q^6)_\infty (q^6; q^6)_\infty (q^2; q^2)_\infty$$
$$= \frac{(-q^2; q^2)_\infty}{(-q^6; q^6)_\infty} (q^6; q^6)_\infty \frac{(q; q^2)_\infty}{(q^3; q^6)_\infty} (q^6; q^6)_\infty (q^2; q^2)_\infty$$
$$= \frac{(q^6; q^6)_\infty}{(-q^6; q^6)_\infty} (q; q)_\infty (-q^2; q^2)_\infty \frac{(q^6; q^6)_\infty}{(q^3; q^6)_\infty}$$
$$(5.24.16) \qquad = \varphi(-q^6)\psi(-q)\psi(q^3).$$

Multiply both sides of (5.24.15) by $f(-q^6)$ and use (5.24.16) with q replaced by $-q^3$ to deduce that

(5.24.17)
$$\varphi(-q^2)\psi(-q)\varphi(-q^{18})\psi(q^3)\psi(-q^9)$$
$$= f^2(-q^3)\psi(q)f(-q^3, -q^{15})f(-q^6) - 2q\varphi(-q^2)\psi(q)\psi(-q^9)\psi(q^9)f(-q^6).$$

From (5.7.6) and (2.14), we find that

(5.24.18)
$$f(-q, -q^5)f(-q^2) = f(-q)\psi(q^3).$$

Using (5.24.18) with q replaced by q^3 in (5.24.17), we find that

(5.24.19)
$$\varphi(-q^2)\psi(-q)\varphi(-q^{18})\psi(q^3)\psi(-q^9)$$
$$= f^3(-q^3)\psi(q)\psi(q^9) - 2q\varphi(-q^2)\psi(q)\psi(-q^9)\psi(q^9)f(-q^6).$$

Using (2.13)–(2.15), or using (2.7)–(2.9), we can easily verify that

(5.24.20)
$$f^3(-q) = \psi(q)\varphi^2(-q).$$

Using (5.13.4) twice with q replaced by $-q$ and $-q^9$, respectively, and (5.24.20) with q replaced by q^3, we deduce from (5.24.19) that

(5.24.21)
$$\varphi(-q)\psi(q)\varphi(-q^9)\psi(q^9)\psi(q^3)$$
$$= \psi(q^3)\varphi^2(-q^3)\psi(q)\psi(q^9) - 2q\varphi(-q^2)\psi(q)\psi(-q^9)\psi(q^9)f(-q^6).$$

Divide both sides of (5.24.21) with $\psi(q)\psi(q^3)\psi(q^9)$ to conclude that

(5.24.22)
$$\varphi(-q)\varphi(-q^9) = \varphi^2(-q^3) - \frac{2q}{\psi(q^3)}\varphi(-q^2)\psi(-q^9)f(-q^6)$$
$$= \varphi^2(-q^3) - 2q\varphi(-q^2)\psi(-q^9)\chi(-q^3),$$

where in the last step we used the extremal equality in (2.14) with q replaced by $-q^3$. If we replace q by $-q$, then (5.24.22) reduces to (5.24.5). Hence, the proof of Entry 3.25 is complete. □

5.25. Proofs of Entries 3.26 and 3.27

The proofs in this section are due to Watson [34].

Recall that we have by (5.19.2),

(5.25.1)
$$\varphi(q) = \varphi(q^4) + 2q\psi(q^8).$$

Returning to (5.20.1), we use (5.25.1) twice. Then we apply Entries 3.2, 3.3, and 3.20, with q replaced by q^4, and Entry 3.21, with q replaced by q^2. In these resulting equalities, we solve for $\varphi(q^4)$, $\varphi(q^{20})$, $\psi(q^8)$, and $\psi(q^{40})$, respectively, and substitute them in the second equality below. Accordingly, we find that

$$G(q) = \frac{\varphi(q) + \varphi(q^5)}{2G(q^4)f(-q^2)}$$
$$= \frac{\varphi(q^4) + \varphi(q^{20})}{2G(q^4)f(-q^2)} + \frac{q\psi(q^8) + q^5\psi(q^{40})}{G(q^4)f(-q^2)}$$
(5.25.2)
$$= \frac{f(-q^8)}{f(-q^2)}\left(G(q^{16}) + qH(-q^4)\right).$$

Performing exactly the same steps on the second equality of (5.20.1), we find that

$$qH(q) = \frac{\varphi(q) - \varphi(q^5)}{2H(q^4)f(-q^2)}$$

$$= \frac{\varphi(q^4) - \varphi(q^{20})}{2H(q^4)f(-q^2)} + \frac{q\psi(q^8) - q^5\psi(q^{40})}{H(q^4)f(-q^2)}$$

(5.25.3)
$$= \frac{f(-q^8)}{f(-q^2)}\left(q^4 H(q^{16}) + qG(-q^4)\right).$$

For brevity, set

(5.25.4) $$T(q) := G(q^{11})H(-q) + q^2 G(-q)H(q^{11}).$$

Next, in the definition (5.25.4), we substitute for each of the functions G and H their respective representations from (5.25.2) and (5.25.3). We therefore deduce that

$$\frac{f(-q^2)}{f(-q^8)} \cdot \frac{f(-q^{22})}{f(-q^{88})} T(q) = \{G(q^{176}) + q^{11}H(-q^{44})\}\{G(-q^4) - q^3 H(q^{16})\}$$

$$+ q^2 \{G(q^{16}) - qH(-q^4)\}\{G(-q^{44}) + q^{33} H(q^{176})\}$$

$$= \{G(-q^4)G(q^{176}) - q^{36}H(-q^4)H(q^{176})\}$$

$$+ q^2 \{G(q^{16})G(-q^{44}) - q^{12}H(q^{16})H(-q^{44})\}$$

$$- q^3 \{G(q^{176})H(q^{16}) - q^{32}G(q^{16})H(q^{176})\}$$

(5.25.5)
$$- q^3 \{G(-q^{44})H(-q^4) - q^8 G(-q^4)H(-q^{44})\}.$$

Recalling the definitions of U and V in (3.31) and (3.32), respectively, recalling the definition (5.25.4), and using Entry 3.4, we find that (5.25.5) can be written in the form

(5.25.6) $$\chi(-q^2)\chi(-q^4)\chi(-q^{22})\chi(-q^{44})T(q) = U(-q^4) + q^2 V(-q^4) - 2q^3.$$

Now replace q by $-q$ in (5.25.6) and subtract the two equalities to deduce that

(5.25.7) $$\chi(-q^2)\chi(-q^4)\chi(-q^{22})\chi(-q^{44})\{T(-q) - T(q)\} = 4q^3.$$

We can obtain a second equation connecting $T(q)$ and $T(-q)$ in the following manner. We record (3.5), (5.25.4), and (3.24), with q replaced by q^{11}. Accordingly,

$$G(q^{11})H(q) - q^2 G(q)H(q^{11}) = 1,$$
$$G(q^{11})H(-q) + q^2 G(-q)H(q^{11}) = T(q),$$
$$G(q^{11})H(-q^{11}) + G(-q^{11})H(q^{11}) = \frac{2}{\chi^2(-q^{22})}.$$

Regarding $G(q^{11})$, $H(q^{11})$, and 1 as the "variables," we conclude from this triple of equations that

(5.25.8) $$\begin{vmatrix} H(q) & -q^2 G(q) & 1 \\ H(-q) & q^2 G(-q) & T(q) \\ H(-q^{11}) & G(-q^{11}) & \dfrac{2}{\chi^2(-q^{22})} \end{vmatrix} = 0.$$

Expanding this determinant (5.25.8) by the last column, using Entry 3.4, recalling the definition (5.25.4), and using Entry 3.20, we find that

$$(5.25.9) \qquad 1 - T(q)T(-q) + \frac{4q^2}{\chi^2(-q^2)\chi^2(-q^{22})} = 0.$$

We now use the theory of modular equations. In Ramanujan's notation for the moduli k and ℓ, let $\alpha = k^2$ and $\beta = \ell^2$, where β is of degree 11 over α. The standard modular equation of degree 11, first found by H. Schröter and rediscovered by Ramanujan [5, p. 363, Entry 7(i)], is given by

$$(5.25.10) \qquad (\alpha\beta)^{1/4} + \{(1-\alpha)(1-\beta)\}^{1/4} + 2^{4/3}\{\alpha\beta(1-\alpha)(1-\beta)\}^{1/12} = 1.$$

We also need the representations [5, p. 124, Entries 12(v), (vii)]
(5.25.11)

$$\chi(q) = 2^{1/6}\left(\frac{q}{\alpha(1-\alpha)}\right)^{1/24} \quad \text{and} \quad \chi(-q^2) = 2^{1/3}\left(\frac{(1-\alpha)q^2}{\alpha^2}\right)^{1/24}.$$

Lastly, we set $-q^2 = Q$. Thus, using (5.25.7), (5.25.9), and (5.25.11), the modular equation (5.25.10), and lastly (5.25.11), we deduce that

(5.25.12)

$$\chi^2(-q^2)\chi^2(-q^4)\chi^2(-q^{22})\chi^2(-q^{44})\{T(q) + T(-q)\}^2$$
$$= 4\chi^2(Q)\chi^2(-Q^2)\chi^2(Q^{11})\chi^2(-Q^{22}) - 16Q\chi^2(-Q^2)\chi^2(-Q^{22}) - 16Q^3$$
$$= 4\chi^2(Q)\chi^2(-Q^2)\chi^2(Q^{11})\chi^2(-Q^{22})$$
$$\times \left(1 - 16\frac{Q}{\chi^2(Q)\chi^2(Q^{11})} - 16\frac{Q^3}{\chi^2(Q)\chi^2(-Q^2)\chi^2(Q^{11})\chi^2(-Q^{22})}\right)$$
$$= 4\chi^2(Q)\chi^2(-Q^2)\chi^2(Q^{11})\chi^2(-Q^{22})\left(1 - 2^{4/3}\{\alpha\beta(1-\alpha)(1-\beta)\}^{1/12} - (\alpha\beta)^{1/4}\right)$$
$$= 4\chi^2(Q)\chi^2(-Q^2)\chi^2(Q^{11})\chi^2(-Q^{22})\{(1-\alpha)(1-\beta)\}^{1/4}$$
$$= 4\chi^2(Q)\chi^2(-Q^2)\chi^2(Q^{11})\chi^2(-Q^{22})\frac{\chi^2(-Q^2)\chi^2(-Q^{22})}{\chi^4(Q)\chi^4(Q^{11})}.$$

Changing back to the original variable q, we take the square root of both sides of (5.25.12) to deduce that

$$(5.25.13) \qquad T(q) + T(-q) = 2\frac{\chi(-q^4)\chi(-q^{44})}{\chi^2(-q^2)\chi^2(-q^{22})} = 2\frac{\chi(q^2)\chi(q^{22})}{\chi(-q^2)\chi(-q^{22})},$$

by (2.15). Now combine (5.25.13) with (5.25.7) to derive the desired formula (3.30).

It remains to prove (3.33) and (3.34). Return to (5.25.6) and insert the just proved formula for $T(q)$ in Entry 3.26. We thus find that

$$(5.25.14) \qquad U(-q^4) + q^2V(-q^4) = \chi(q^2)\chi(-q^4)\chi(q^{22})\chi(-q^{44}).$$

Changing the sign of q^2 in (5.25.14), we find that

$$(5.25.15) \qquad U(-q^4) - q^2V(-q^4) = \chi(-q^2)\chi(-q^4)\chi(-q^{22})\chi(-q^{44}).$$

Multiplying (5.25.14) and (5.25.15) together, we arrive at

$$U^2(-q^4) - q^4V^2(-q^4) = \chi(q^2)\chi(-q^2)\chi(q^{22})\chi(-q^{22})\chi^2(-q^4)\chi^2(-q^{44})$$
$$(5.25.16) \qquad\qquad = \chi^3(-q^4)\chi^3(-q^{44}),$$

by (2.15). If we replace $-q^4$ by q in (5.25.16), we obtain (3.33).

Finally, we prove (3.34). We record the definition (3.31) of $U(q)$, (3.5) with q replaced by q^4, and (3.3) with q replaced by q^{11}, in the array

$$G(q)G(q^{44}) + q^9 H(q)H(q^{44}) = U(q),$$
$$H(q^4)G(q^{44}) - q^8 G(q^4)H(q^{44}) = 1,$$
$$G(q^{11})G(q^{44}) + q^{11} H(q^{11})H(q^{44}) = \chi^2(q^{11}).$$

Regard this system of equations as three equations in the unknowns $G(q^{44})$, $q^8 H(q^{44})$, and -1. It follows that

(5.25.17)
$$\begin{vmatrix} G(q) & qH(q) & U(q) \\ H(q^4) & -G(q^4) & 1 \\ G(q^{11}) & q^3 H(q^{11}) & \chi^2(q^{11}) \end{vmatrix} = 0.$$

Expanding the determinant (5.25.17) by the last column, and then using the definition (3.32) of V, (3.5), and (3.3), we find that

$$U(q)V(q) + q - \chi^2(q^{11})\chi^2(q) = 0,$$

which is precisely (3.34).

5.26. Proof of the First Part of Entry 3.28

Our argument below is the same as that of Bressoud [14].

To prove (3.35), we use (2.11) to rewrite the identity as

(5.26.1)
$$\frac{f(-q^{34}, -q^{51})f(-q^2, -q^8) - q^3 f(-q^4, -q^6)f(-q^{17}, -q^{68})}{f(-q^2, -q^3)f(-q^{68}, -q^{102}) + q^7 f(-q, -q^4)f(-q^{34}, -q^{136})}$$
$$= \frac{\chi(-q)f(-q^2)f(-q^{17})}{\chi(-q^{17})f(-q)f(-q^{34})}.$$

From (2.9) and (2.10), it is easy to see that

$$\chi(-q)f(-q^2)f(-q^{17}) = (q;q^2)_\infty (q^2;q^2)_\infty (q^{17};q^{17})_\infty$$
$$= (q;q)_\infty (q^{17};q^{34})_\infty (q^{34};q^{34})_\infty$$
$$= f(-q)\chi(-q^{17})f(-q^{34}).$$

Thus, the right-hand side of (5.26.1) equals 1, and so (3.35) is equivalent to

(5.26.2)
$$f(-q^{34}, -q^{51})f(-q^2, -q^8) - q^3 f(-q^4, -q^6)f(-q^{17}, -q^{68})$$
$$= f(-q^2, -q^3)f(-q^{68}, -q^{102}) + q^7 f(-q, -q^4)f(-q^{34}, -q^{136}).$$

We now apply (4.16) with $\alpha = 1$ and $\beta = \frac{17}{2}$ to obtain

(5.26.3) $$\sum_{k=1}^{2} F(1, \tfrac{17}{2}, 3, 5, \tfrac{7}{2}, k)$$
$$= q^{\frac{7}{8}}(f(-q^{34}, -q^{51})f(-q^2, -q^8) - q^3 f(-q^4, -q^6)f(-q^{17}, -q^{68})).$$

Similarly, letting $\alpha = \frac{1}{2}$ and $\beta = 17$ in (4.15), we deduce that

$$\sum_{k=1}^{2} F(\tfrac{1}{2}, 17, 1, 5, \tfrac{7}{2}, k) \tag{5.26.4}$$
$$= q^{\frac{7}{8}} (f(-q^2, -q^3)f(-q^{68}, -q^{102}) + q^7 f(-q, -q^4)f(-q^{34}, -q^{136})).$$

The two sets of parameters $\{1, \frac{17}{2}, 3, 5, \frac{7}{2}\}$ and $\{\frac{1}{2}, 17, 1, 5, \frac{7}{2}\}$ give rise to the same series on the right-hand side of (4.8), since the parameters satisfy the conditions in (4.9). Hence, the right-hand sides of (5.26.3) and (5.26.4) are equal. This completes the proof of (5.26.2) and so also of the first part of Entry 3.28.

5.27. Proof of the Equivalence of Entries 3.31 and 3.32

For brevity, define

$$M(q) := G(q^2)G(q^{33}) + q^7 H(q^2)H(q^{33}), \tag{5.27.1}$$

$$N(q) := G(q^{66})H(q) - q^{13}H(q^{66})G(q), \tag{5.27.2}$$

$$R(q) := G(q^{66})H(q^{11}) - q^{11}H(q^{66})G(q^{11})$$

$$T(q) := G(q^3)G(q^{22}) + q^5 H(q^3)H(q^{22}), \tag{5.27.3}$$

$$U(q) := G(q^{11})H(q^6) - qH(q^{11})G(q^6). \tag{5.27.4}$$

Using (5.27.2), Entry 3.4 with q replaced by q^6, and Entry 3.8 with q replaced by q^{11}, we consider the system of three equations,

$$N(q) = G(q^{66})H(q) - q^{13}H(q^{66})G(q),$$
$$1 = G(q^{66})H(q^6) - q^{12}H(q^{66})G(q^6),$$
$$\frac{\chi(-q^{11})}{\chi(-q^{33})} = R(q) = G(q^{66})H(q^{11}) - q^{11}H(q^{66})G(q^{11}). \tag{5.27.5}$$

It follows that

$$\begin{vmatrix} H(q) & -q^{13}G(q) & N(q) \\ H(q^6) & -q^{12}G(q^6) & 1 \\ H(q^{11}) & -q^{11}G(q^{11}) & R(q) \end{vmatrix} = 0.$$

Expanding this determinant along the last column and using (5.27.4), Entry 3.4, (5.27.5), and Entry 3.8, we deduce that

(5.27.6)

$$0 = N(q)\left(-q^{11}H(q^6)G(q^{11}) + q^{12}G(q^6)H(q^{11})\right) + q^{11}H(q)G(q^{11}) - q^{13}G(q)H(q^{11})$$
$$+ R(q)\left(-q^{12}G(q^6)H(q) + q^{13}G(q)H(q^6)\right)$$
$$= -N(q)q^{11}U(q) + q^{11} - q^{12}\frac{\chi(-q^{11})}{\chi(-q^{33})}\frac{\chi(-q)}{\chi(-q^3)}.$$

Hence, if we define

$$W(q) := \frac{\chi(-q)\chi(-q^{11})}{\chi(-q^3)\chi(-q^{33})}, \tag{5.27.7}$$

then, from (5.27.6), we deduce that

$$N(q)U(q) = 1 - qW(q). \tag{5.27.8}$$

Next, using (5.27.1), Entry 3.4 with q replaced by q^3, and Entry 3.7 with q replaced by q^{11}, we consider the system of equations

$$M(q) = G(q^2)G(q^{33}) + q^7 H(q^2)H(q^{33}),$$
$$1 = H(q^3)G(q^{33}) - q^6 G(q^3)H(q^{33}),$$

(5.27.9) $\quad \dfrac{\chi(-q^{33})}{\chi(-q^{11})} =: S(q) = G(q^{22})G(q^{33}) + q^{11} H(q^{22})H(q^{33}).$

It follows that

$$\begin{vmatrix} G(q^2) & q^7 H(q^2) & M(q) \\ H(q^3) & -q^6 G(q^3) & 1 \\ G(q^{22}) & q^{11} H(q^{22}) & S(q) \end{vmatrix} = 0.$$

Expanding the determinant above along the last column and employing (5.27.3), Entry 3.4 with q replaced by q^2, (5.27.9), and Entry 3.7, we find that

(5.27.10)
$$0 = M(q)\left(q^{11} H(q^3)H(q^{22}) + q^6 G(q^3)G(q^{22})\right) - q^{11} G(q^2)H(q^{22}) + q^7 H(q^2)G(q^{22})$$
$$+ S(q)\left(-q^6 G(q^2)G(q^3) - q^7 H(q^2)H(q^3)\right)$$
$$= M(q)q^6 T(q) + q^7 - q^6 \dfrac{\chi(-q^{33})}{\chi(-q^{11})} \dfrac{\chi(-q^3)}{\chi(-q)}.$$

Hence, using the definition of $W(q)$ in (5.27.7), we deduce from (5.27.10) that

(5.27.11) $\quad M(q)T(q) = -q + \dfrac{1}{W(q)}.$

Hence, dividing (5.27.11) by (5.27.8), we conclude that

(5.27.12) $\quad \dfrac{M(q)T(q)}{N(q)U(q)} = \dfrac{1}{W(q)}.$

Examining Entries 3.31 and 3.32, we see that it suffices to prove just one of them, for then the other one would follow immediately from (5.27.12).

5.28. Proof of Entry 3.33

We provide two proofs.

First Proof of Entry 3.33. Let us define $K(q)$ and $L(q)$ by

(5.28.1) $\quad K(q) := G(q)G(q^{54}) + q^{11} H(q)H(q^{54}),$
(5.28.2) $\quad L(q) := H(q^2)G(q^{27}) - q^5 G(q^2)H(q^{27}),$

so that Entry 3.33 reads

(5.28.3) $\quad \dfrac{K(q)}{L(q)} = \dfrac{\chi(-q^3)\chi(-q^{27})}{\chi(-q)\chi(-q^9)}.$

Starting from (3.15) and arguing as in (4.32), we find that

(5.28.4) $\quad \dfrac{\chi(-q^6)\chi(-q^9)}{\chi(-q^2)\chi(-q^3)} G(-q) + \dfrac{\chi(-q^6)\chi(q^9)}{\chi(-q^2)\chi(q^3)} G(q) = 2 \dfrac{G(q^{36})}{\chi^2(-q^2)}.$

By (2.15), we see that (5.28.4) simplifies to

(5.28.5) $\quad \chi(q^3)\chi(-q^9)G(-q) + \chi(-q^3)\chi(q^9)G(q) = 2 \dfrac{G(q^{36})}{\chi(-q^2)}.$

Similarly, we find that

(5.28.6) $$\chi(-q^3)\chi(q^9)H(q) - \chi(q^3)\chi(-q^9)H(-q) = 2q^7\frac{H(q^{36})}{\chi(-q^2)}.$$

In (5.28.1), we replace q by q^2 and employ (5.28.5) and (5.28.6) with q replaced by q^3 to find that

$$2\frac{K(q^2)}{\chi(-q^6)} = \frac{2}{\chi(-q^6)}\left\{G(q^2)G(q^{108}) + q^{22}H(q^2)H(q^{108})\right\}$$
$$= G(q^2)\left\{\chi(q^9)\chi(-q^{27})G(-q^3) + \chi(-q^9)\chi(q^{27})G(q^3)\right\}$$
$$+ qH(q^2)\left\{\chi(-q^9)\chi(q^{27})H(q^3) - \chi(q^9)\chi(-q^{27})H(-q^3)\right\}$$
$$= \chi(q^9)\chi(-q^{27})\left\{G(q^2)G(-q^3) - qH(q^2)H(-q^3)\right\}$$
(5.28.7) $$+ \chi(-q^9)\chi(q^{27})\left\{G(q^2)G(q^3) + qH(q^2)H(q^3)\right\}.$$

Using (3.8) twice, once with q replaced by $-q$, we see that (5.28.7) can be put in the form

$$2\frac{K(q^2)}{\chi(-q^6)} = \chi(q^9)\chi(-q^{27})\frac{\chi(q^3)}{\chi(q)} + \chi(-q^9)\chi(q^{27})\frac{\chi(-q^3)}{\chi(-q)}.$$

Using (2.15), we conclude that
(5.28.8)
$$2K(q^2) = \frac{\chi(-q^6)}{\chi(-q^2)}\left\{\chi(-q)\chi(q^3)\chi(q^9)\chi(-q^{27}) + \chi(q)\chi(-q^3)\chi(-q^9)\chi(q^{27})\right\}.$$

To obtain the desired expression for $L(q^2)$, we use Lemma 4.3. Then, in (5.28.2), replacing q by q^2, employing (4.36) and (4.37), and arguing as in (5.28.7), we find that

$$2\frac{L(q^2)}{\chi(-q^{18})} = \chi(q)\chi(-q^3)\left\{G(q^{54})H(q^9) - q^9H(q^{54})G(q^9)\right\}$$
(5.28.9) $$+ \chi(-q)\chi(q^3)\left\{G(q^{54})H(-q^9) + q^9H(q^{54})G(-q^9)\right\}.$$

Using (3.9), with q replaced by q^9 and $-q^9$, respectively, we find from (5.28.9) that

$$2\frac{L(q^2)}{\chi(-q^{18})} = \chi(q)\chi(-q^3)\frac{\chi(-q^9)}{\chi(-q^{27})} + \chi(-q)\chi(q^3)\frac{\chi(q^9)}{\chi(q^{27})},$$

which, by (2.15), implies that
(5.28.10)
$$2L(q^2) = \frac{\chi(-q^{18})}{\chi(-q^{54})}\left\{\chi(q)\chi(-q^3)\chi(-q^9)\chi(q^{27}) + \chi(-q)\chi(q^3)\chi(q^9)\chi(-q^{27})\right\}.$$

Dividing (5.28.8) by (5.28.10), we obtain (5.28.3) with q replaced by q^2. Hence, the proof of Entry 3.33 is complete. □

Second Proof of Entry 3.33. Recall once more the definitions

$$g(q) = f(-q)G(q) = f(-q^2, -q^3) \quad \text{and} \quad h(q) = f(-q)H(q) = f(-q, -q^4).$$

Our proof of Entry 3.33 uses Entries 3.6, 3.7, and 3.8, which we write in their equivalent forms (see (5.6.1), (5.7.3), and (5.8.2)),

(5.28.11) $$g(q)g(q^9) + q^2 h(q)h(q^9) = f^2(-q^3),$$

(5.28.12) $$g(q^2)g(q^3) + qh(q^2)h(q^3) = \psi(q)\varphi(-q^3),$$

(5.28.13) $$g(q^6)h(q) - qg(q)h(q^6) = \psi(q^3)\varphi(-q).$$

Let us define $M(q)$ and $N(q)$ by

(5.28.14) $$M(q) := h(q^2)g(q^{27}) - q^5 g(q^2)h(q^{27}),$$

(5.28.15) $$N(q) := g(q)g(q^{54}) + q^{11} h(q)h(q^{54}).$$

By (2.11) and (2.14), Entry 3.33 is equivalent to the identity

(5.28.16) $$\frac{N(q)}{M(q)} = \frac{f(-q)f(-q^{54})\chi(-q^3)\chi(-q^{27})}{f(-q^2)f(-q^{27})\chi(-q)\chi(-q^9)} = \frac{\chi(-q^3)}{\chi(-q^9)}.$$

By (5.28.15), (5.28.11), and (5.28.13) with q replaced by q^6 and q^9, respectively, in the latter two cases, we deduce the following system of three equations:

$$g(q)g(q^{54}) + q^{11} h(q)h(q^{54}) = N(q),$$
$$g(q^6)g(q^{54}) + q^{12} h(q^6)h(q^{54}) = f^2(-q^{18}),$$
$$h(q^9)g(q^{54}) - q^9 g(q^9)h(q^{54}) = \psi(q^{27})\varphi(-q^9).$$

Regarding this system in the variables $G(q^{54})$, $q^9 h(q^{54})$, and -1, we find that

(5.28.17) $$\begin{vmatrix} g(q) & q^2 h(q) & N(q) \\ g(q^6) & q^3 h(q^6) & f^2(-q^{18}) \\ h(q^9) & -g(q^9) & \psi(q^{27})\varphi(-q^9) \end{vmatrix} = 0.$$

Expanding the determinant in (5.28.17) along the last column, we find that

(5.28.18)
$$-N(q)\left\{g(q^6)g(q^9) + q^3 h(q^6)h(q^9)\right\}$$
$$+ f^2(-q^{18})\left\{g(q)g(q^9) + q^2 h(q)h(q^9)\right\}$$
$$- q^2 \varphi(-q^9)\psi(q^{27})\left\{h(q)g(q^6) - qg(q)h(q^6)\right\} = 0.$$

Using (5.28.12) with q replaced by q^3, (5.28.11), and (5.28.13) in (5.28.18), we find that

$$-N(q)\psi(q^3)\varphi(-q^9) + f^2(-q^{18})f^2(-q^3) - q^2 \psi(q^{27})\varphi(-q^9)\psi(q^3)\varphi(-q) = 0.$$

Solving for $N(q)$, we deduce that

(5.28.19) $$N(q) = \frac{f^2(-q^{18})f^2(-q^3)}{\psi(q^3)\varphi(-q^9)} - q^2 \psi(q^{27})\varphi(-q).$$

Next, we determine $M(q)$. By (5.28.14), (5.28.12), and (5.28.11) with q replaced by q^9 and q^3, respectively, in the latter two equalities, we find that

$$h(q^2)g(q^{27}) - q^5 g(q^2)h(q^{27}) = M(q),$$
$$g(q^{18})g(q^{27}) + q^9 h(q^{18})h(q^{27}) = \psi(q^9)\varphi(-q^{27}),$$
$$g(q^3)g(q^{27}) + q^6 h(q^3)h(q^{27}) = f^2(-q^9).$$

Regarding this system in the variables $g(q^{27})$, $q^5 h(q^{27})$, and -1 and arguing as we did above, we find that

$$qM(q)\psi(q^9)\varphi(-q^3) - \psi(q^9)\varphi(-q^{27})\psi(q)\varphi(-q^3) + f^2(-q^9)f^2(-q^6) = 0.$$

Solving for $qM(q)$, we arrive at

(5.28.20) $$qM(q) = \psi(q)\varphi(-q^{27}) - \frac{f^2(-q^9)f^2(-q^6)}{\psi(q^9)\varphi(-q^3)}.$$

Recall that, by (5.7.6) and (5.7.8),

(5.28.21) $$f(-q, -q^5) = \chi(-q)\psi(q^3)$$

and

(5.28.22) $$f(q, q^2) = \frac{\varphi(-q^3)}{\chi(-q)}.$$

By (2.19) with $n = 3$ (see also [**5**, p. 49, Cor.]), we find that

(5.28.23) $$\varphi(-q) = \varphi(-q^9) - 2qf(-q^3, -q^{15}),$$
(5.28.24) $$\psi(q) = f(q^3, q^6) + q\psi(q^9).$$

Using (5.28.21) and (5.28.22) in (5.28.23) and (5.28.24) with q replaced by q^3, we obtain, respectively,

(5.28.25) $$\varphi(-q) = \varphi(-q^9) - 2q\chi(-q^3)\psi(q^9),$$

(5.28.26) $$\psi(q) = \frac{\varphi(-q^9)}{\chi(-q^3)} + q\psi(q^9).$$

We deduce from (5.28.25) and (5.28.26) that

(5.28.27) $$\chi(-q^3)\psi(q^9) = \frac{1}{2q}\left\{\varphi(-q^9) - \varphi(-q)\right\},$$

(5.28.28) $$\psi(q)\chi(-q^3) = \frac{1}{2}\left\{3\varphi(-q^9) - \varphi(-q)\right\}.$$

By (2.14), we easily find that

(5.28.29) $$\frac{f^2(-q^2)\chi(-q)}{\varphi(-q)} = \psi(q),$$

(5.28.30) $$\frac{f^2(-q)}{\psi(q)} = \varphi(-q)\chi(-q).$$

By (5.28.19), (5.28.29), and (5.28.30) with q replaced by q^9 and q^3, respectively, in the latter two equations, we find that

(5.28.31) $$\chi(-q^9)N(q) = \chi(-q^3)\varphi(-q^3)\psi(q^9) - q^2\psi(q^{27})\chi(-q^9)\varphi(-q).$$

Using (5.28.27) twice in (5.28.31) with q replaced by q and q^3, respectively, we find that

$$\chi(-q^9)N(q) = \frac{1}{2q}\varphi(-q^3)\left\{\varphi(-q^9) - \varphi(-q)\right\} - q^2\frac{1}{2q^3}\left\{\varphi(-q^{27}) - \varphi(-q^3)\right\}\varphi(-q)$$

(5.28.32) $$= \frac{1}{2q}\left\{\varphi(-q^3)\varphi(-q^9) - \varphi(-q)\varphi(-q^{27})\right\}.$$

Similarly, by (5.28.20), (5.28.29), and (5.28.30) with q replaced by q^3 and q^9, respectively, we find that

(5.28.33) $\qquad q\chi(-q^3)M(q) = \chi(-q^3)\varphi(-q^{27})\psi(q) - \psi(q^3)\chi(-q^9)\varphi(-q^9).$

Using (5.28.28) twice with q replaced by q and q^3, respectively, we find that

$$q\chi(-q^3)M(q) = \frac{1}{2}\varphi(-q^{27})\left\{3\varphi(-q^9) - \varphi(-q)\right\} - \frac{1}{2}\left\{3\varphi(-q^{27}) - \varphi(-q^3)\right\}\varphi(-q^9)$$

(5.28.34) $\qquad = \frac{1}{2}\left\{\varphi(-q^3)\varphi(-q^9) - \varphi(-q)\varphi(-q^{27})\right\}.$

Dividing (5.28.34) by (5.28.32), we see that (5.28.16) is verified. Hence, the second proof of Entry 3.33 is complete. $\qquad\square$

5.29. Proof of Entry 3.34

Our proof is a moderate modification of the proof given by Bressoud [14].

Using (1.2), (2.6), and some elementary product manipulations, we can show that

(5.29.1) $\qquad G(q)G(-q) = \dfrac{f(q^4, q^6)}{f(-q^2)} \quad\text{and}\quad H(q)H(-q) = \dfrac{f(q^2, q^8)}{f(-q^2)}.$

Adding Entries 3.20 and 3.21, we find that

(5.29.2) $\qquad G(q)H(-q) = \dfrac{1}{f(-q^2)}\left\{\psi(q^2) + q\psi(q^{10})\right\}.$

Next, we recall Entry 3.34:

$$\left\{G(q)G(-q^{19}) - q^4 H(q)H(-q^{19})\right\}\left\{G(-q)G(q^{19}) - q^4 H(-q)H(q^{19})\right\}$$
(5.29.3) $\qquad = G(q^2)G(q^{38}) + q^8 H(q^2)H(q^{38}).$

Expanding the product on the left side of (5.29.3) and then using (5.29.1) and (5.29.2), we find that

(5.29.4)
$$\left\{G(q)G(-q^{19}) - q^4 H(q)H(-q^{19})\right\}\left\{G(-q)G(q^{19}) - q^4 H(-q)H(q^{19})\right\}$$
$$= G(q)G(-q)G(q^{19})G(-q^{19}) + q^8 H(q)H(-q)H(q^{19})H(-q^{19})$$
$$\quad - q^4 G(-q)H(q)G(q^{19})H(-q^{19}) - q^4 G(q)H(-q)G(-q^{19})H(q^{19})$$
$$= \frac{1}{f(-q^2)f(-q^{38})} f(q^4, q^6) f(q^{76}, q^{114}) + q^8 \frac{1}{f(-q^2)f(-q^{38})} f(q^2, q^8) f(q^{38}, q^{152})$$
$$\quad - q^4 \frac{1}{f(-q^2)f(-q^{38})}\left\{\psi(q^2) - q\psi(q^{10})\right\}\left\{\psi(q^{38}) + q^{19}\psi(q^{190})\right\}$$
$$\quad - q^4 \frac{1}{f(-q^2)f(-q^{38})}\left\{\psi(q^2) + q\psi(q^{10})\right\}\left\{\psi(q^{38}) - q^{19}\psi(q^{190})\right\}$$
$$= \frac{1}{f(-q^2)f(-q^{38})}\left\{f(q^4, q^6)f(q^{76}, q^{114}) + q^8 f(q^2, q^8)f(q^{38}, q^{152})\right.$$
$$\left. \quad - 2q^4 \psi(q^2)\psi(q^{38}) + 2q^{24}\psi(q^{10})\psi(q^{190})\right\}.$$

Using [**21**, Cor. 1.3], we readily find that

$$\frac{1}{(q^4;q^{10})_\infty(q^6;q^{10})_\infty} = \left(\frac{2}{5+\sqrt{5}}\right)^{1/2} \exp\left(\frac{\pi^2}{30t} + \frac{11t}{30}\right)$$
(6.1.7)
$$\times \frac{1}{(e(3/5)u^{1/10};u^{1/10})_\infty(e(2/5)u^{1/10};u^{1/10})_\infty},$$

where $u = e^{-4\pi^2/t}$. For the second product in (6.1.6), we apply the last part of Corollary 1.3 in [**21**], with q replaced by q^{10}, $\alpha = \frac{1}{10}$, and $\mu = \frac{1}{2}$, to find that

(6.1.8) $\quad \dfrac{1}{(-q;q^{10})_\infty(-q^9;q^{10})_\infty}$

$$= \exp\left(-\frac{\pi^2}{60t} - \frac{23t}{60}\right) \frac{1}{(e(9/10)u^{1/20};u^{1/20})_\infty(e(1/10)u^{1/20};u^{1/20})_\infty},$$

where $u = e^{-4\pi^2/t}$. Substituting (6.1.7) and (6.1.8) in (6.1.6), we deduce that

(6.1.9) $\quad G(-q) = \left(\dfrac{2}{5+\sqrt{5}}\right)^{1/2} \exp\left(\dfrac{\pi^2}{60t} - \dfrac{t}{60}\right) \dfrac{1}{(e(3/5)u^{1/10};u^{1/10})_\infty}$

$$\times \frac{1}{(e(2/5)u^{1/10};u^{1/10})_\infty(e(9/10)u^{1/20};u^{1/20})_\infty(e(1/10)u^{1/20};u^{1/20})_\infty}.$$

Second, from (1.2),

(6.1.10) $\quad H(-q) = \dfrac{1}{(q^2;q^{10})_\infty(q^8;q^{10})_\infty(-q^3;q^{10})_\infty(-q^7;q^{10})_\infty}.$

Using again [**21**, Cor. 1.3], we find that

$$\frac{1}{(q^2;q^{10})_\infty(q^8;q^{10})_\infty} = \left(\frac{2}{5-\sqrt{5}}\right)^{1/2} \exp\left(\frac{\pi^2}{30t} - \frac{t}{30}\right)$$
(6.1.11)
$$\times \frac{1}{(e(4/5)u^{1/10};u^{1/10})_\infty(e(1/5)u^{1/10};u^{1/10})_\infty},$$

where $u = e^{-4\pi^2/t}$. For the second product in (6.1.10), we again apply the last part of Corollary 1.3 in [**21**], with q replaced by q^{10}, $\alpha = \frac{3}{10}$, and $\mu = \frac{1}{2}$, to find that

(6.1.12) $\quad \dfrac{1}{(-q^3;q^{10})_\infty(-q^7;q^{10})_\infty}$

$$= \exp\left(-\frac{\pi^2}{60t} + \frac{13t}{60}\right) \frac{1}{(e(7/10)u^{1/20};u^{1/20})_\infty(e(3/10)u^{1/20};u^{1/20})_\infty},$$

where $u = e^{-4\pi^2/t}$. Combining (6.1.11) and (6.1.12) with (6.1.10), we deduce that

(6.1.13) $\quad H(-q) = \left(\dfrac{2}{5-\sqrt{5}}\right)^{1/2} \exp\left(\dfrac{\pi^2}{60t} + \dfrac{11t}{60}\right) \dfrac{1}{(e(4/5)u^{1/10};u^{1/10})_\infty}$

$$\frac{1}{(e(1/5)u^{1/10};u^{1/10})_\infty(e(7/10)u^{1/20};u^{1/20})_\infty(e(3/10)u^{1/20};u^{1/20})_\infty}.$$

The familiar transformation formula for the Dedekind eta-function can be written in the form [**21**, Cor. 1.2]

(6.1.14) $\qquad (q;q)_\infty = \sqrt{\dfrac{2\pi}{t}} \exp\left(-\dfrac{\pi^2}{6t} + \dfrac{t}{24}\right) (u;u)_\infty,$

where $q = e^{-t}, t > 0$, and $u = e^{-4\pi^2/t}$. Hence, from the definition (2.10) of $\chi(q)$, we find that

$$(6.1.15) \qquad \chi(-q) = \sqrt{2}\exp\left(-\frac{\pi^2}{12t} - \frac{t}{24}\right)\frac{(u;u)_\infty}{(\sqrt{u};\sqrt{u})_\infty}.$$

We need to explicitly determine several coefficients in the expansion of the quotient on the right hand side of (6.1.15). Thus, letting $v = \sqrt{u}$ and using *Mathematica*, we find that

$$(6.1.16) \qquad \chi(-q) = \sqrt{2}\exp\left(-\frac{\pi^2}{12t} - \frac{t}{24}\right)(1 + v + v^2 + 2v^3 + 2v^4 + 3v^5 + 4v^6 \\ + 5v^7 + 6v^8 + 8v^9 + 10v^{10} + O(v^{11}))$$

and

$$(6.1.17) \qquad \frac{1}{\chi(-q)} = \frac{1}{\sqrt{2}}\exp\left(\frac{\pi^2}{12t} + \frac{t}{24}\right)(1 - v - v^3 + v^4 - v^5 + v^6 \\ - v^7 + 2v^8 - 2v^9 + 2v^{10} + O(v^{11})),$$

as v (or t) tends to 0.

We are now ready to provide asymptotic "proofs" of the remaining five entries. For the proofs of Entries 3.30 and 3.31, we show that both sides of each identity have the same exponentially asymptotic behavior and calculate only one or two of the secondary terms arising from the expansions of the products on the rights sides of (6.1.1), (6.1.2), (6.1.9), and (6.1.13). However, in order to give convincing arguments for the "proofs" of Entries 3.28, 3.29, and 3.35, it will be necessary to calculate several terms of the aforementioned expansions. Corresponding calculations of the asymptotic series in Entries 3.30 and 3.31 are also much easier than they are for the other three entries.

6.2. "Proof" of Entry 3.30

First, from (6.1.16) and (6.1.17), we find that

$$(6.2.1) \qquad \frac{\chi(-q^2)}{\chi(-q^{38})} = \exp\left(-\frac{3\pi^2}{76t} + \frac{3t}{2}\right)\left(1 - e^{-\pi^2/(19t)} + \cdots\right).$$

Second, from (6.1.1), (6.1.2), (6.1.4), and (6.1.5),

$$G(q^{19})H(q^4) - q^3 G(q^4)H(q^{19})$$
$$= \exp\left(\frac{23\pi^2}{15 \cdot 76t} + \frac{5t}{12}\right)\left(e^{-4\pi^2/(5\cdot 19t)} + e^{-44\pi^2/(5\cdot 19t)} + \cdots\right)$$
$$(6.2.2) \qquad = \exp\left(-\frac{5\pi^2}{3 \cdot 76t} + \frac{5t}{12}\right)\left(1 + e^{-40\pi^2/(5\cdot 19t)} + \cdots\right).$$

Third, from (6.1.1), (6.1.2), (6.1.4), (6.1.5), (6.1.9), and (6.1.13),
$$(6.2.3)$$
$$G(q^{76})H(-q) + q^{15}G(-q)H(q^{76}) = \exp\left(\frac{\pi^2}{3\cdot 19t} - \frac{13t}{12}\right)\left(1 + e^{-\pi^2/(19t)} + \cdots\right).$$

Combining (6.2.2) and (6.2.3), we find that
$$(6.2.4)$$
$$\frac{G(q^{19})H(q^4) - q^3 G(q^4)H(q^{19})}{G(q^{76})H(-q) + q^{15}G(-q)H(q^{76})} = \exp\left(-\frac{3\pi^2}{76t} + \frac{3t}{2}\right)\left(1 - e^{-\pi^2/(19t)} + \cdots\right).$$

6.3. "Proof" of Entry 3.31

From (6.1.16) and (6.1.17), we find that

(6.3.1) $$\frac{\chi(-q^3)}{\chi(-q^{11})} = \exp\left(-\frac{2\pi^2}{3\cdot 33t} + \frac{t}{3}\right)(1+\cdots).$$

On the other hand, by (6.1.1), (6.1.2), (6.1.4), and (6.1.5),

(6.3.2) $$G(q^2)G(q^{33}) + q^7 H(q^2)H(q^{33}) = \exp\left(\frac{7\pi^2}{6\cdot 33t} - \frac{7t}{12}\right)(1+\cdots)$$

and

(6.3.3) $$G(q^{66})H(q) - q^{13}H(q^{66})G(q) = \exp\left(\frac{67\pi^2}{15\cdot 66t} - \frac{11t}{12}\right)\left(e^{-4\pi^2/(5\cdot 66t)} + \cdots\right)$$
$$= \exp\left(\frac{11\pi^2}{3\cdot 66t} - \frac{11t}{12}\right)(1+\cdots).$$

Combining (6.3.2) and (6.3.3), we deduce that

(6.3.4) $$\frac{G(q^2)G(q^{33}) + q^7 H(q^2)H(q^{33})}{G(q^{66})H(q) - q^{13}H(q^{66})G(q)} = \exp\left(-\frac{2\pi^2}{3\cdot 33t} + \frac{t}{3}\right)(1+\cdots).$$

Asymptotically, the left and right sides of (6.3.1) and (6.3.4) are the same.

6.4. "Proof" of the Second Part of Entry 3.28

To consider the influence of each term in (3.36), we need to calculate several terms of each asymptotic expansion. Roughly speaking, if a is much larger than b, in the asymptotic series for $G(q^a)$ ($H(q^a)$) and $G(q^b)$ ($H(q^b)$), the terms of $G(q^a)$ ($H(q^a)$) are much larger than those of $G(q^b)$ ($H(q^b)$). However, in the calculations which follow, we cannot ignore the terms for $G(q^2)$ and $H(q^2)$ in relation to those for $G(q^{17})$ and $H(q^{17})$. More precisely, in order to determine the necessary terms in the asymptotic expansions below, we need the first two terms in each of the asymptotic series of $G(q^2)$ and $H(q^2)$.

First, if $q = e^{-t}$ and $v = e^{-2\pi^2/(17t)}$, using (6.1.16) and (6.1.17), we find that

(6.4.1)
$$\frac{\chi^3(-q^{17})}{\chi^3(-q)} + 2q^2 + q^4\frac{\chi^3(-q)}{\chi^3(-q^{17})} = e^{-2t}\left(v^{-2}\left(1 + v + v^2 + 2v^3 + \cdots\right)^3\right.$$
$$\left.+ 2 + v^2\left(1 - v - v^3 + \cdots\right)^3\right)$$
$$= e^{-2t}\left(v^{-2}\left(1 + 3v + 6v^2 + 13v^3 + 24v^4 + 42v^5 + 73v^6 + 120v^7 + O(v^8)\right)\right.$$
$$\left.+ 2 + v^2\left(1 - 3v + 3v^2 - 4v^3 + O(v^4)\right)\right)$$
$$= e^{-2t}\left(v^{-2} + 3v^{-1} + 8 + 13v + 25v^2 + 39v^3 + 76v^4 + 124v^5 + O(v^6)\right).$$

Second, let A_n and B_n denote the products on the right sides of (6.1.1) and (6.1.2), respectively, when q is replaced by q^n. We then find from (6.1.1) and (6.1.2)

that
(6.4.2)
$$U = G(q^{17})H(q^2) - q^3 G(q^2)H(q^{17}) = \frac{1}{\sqrt{5}} \exp\left(\frac{19\pi^2}{15 \cdot 34 t} + \frac{t}{12}\right)(A_{17}B_2 - A_2 B_{17})$$

and
$$V = G(q)G(q^{34}) + q^7 H(q)H(q^{34}) = \exp\left(\frac{7\pi^2}{3 \cdot 34 t} - \frac{7t}{12}\right)$$
(6.4.3)
$$\times \left(\frac{2}{5 - \sqrt{5}} A_1 A_{34} + \frac{2}{5 + \sqrt{5}} B_1 B_{34}\right).$$

We thus see from the second part of Entry 3.28 and (6.4.2) and (6.4.3) that we need to calculate the asymptotic expansions for

(6.4.4) $U^4 V^4$
$$= \frac{1}{25} \exp\left(\frac{36\pi^2}{5 \cdot 17 t} - 2t\right)(A_{17}B_2 - A_2 B_{17})^4 \left(\frac{2}{5 - \sqrt{5}} A_1 A_{34} + \frac{2}{5 + \sqrt{5}} B_1 B_{34}\right)^4$$

and

(6.4.5) $qU^2 V^2$
$$= \frac{1}{5} \exp\left(\frac{18\pi^2}{5 \cdot 17 t} - 2t\right)(A_{17}B_2 - A_2 B_{17})^2 \left(\frac{2}{5 - \sqrt{5}} A_1 A_{34} + \frac{2}{5 + \sqrt{5}} B_1 B_{34}\right)^2.$$

Letting $v = Q^{5/2}$, using (6.1.1), (6.1.2), (6.1.4), and (6.1.5), and employing *Mathematica*, we find that
(6.4.6)
$$(A_{17}B_2 - A_2 B_{17})^4 = 25\left(Q^4 - 4Q^{23/2} + 4Q^{14} - 4Q^{33/2} + 10Q^{19} + O(Q^{43/2})\right),$$

(6.4.7) $\quad (A_{17}B_2 - A_2 B_{17})^2 = 5\left(Q^2 - 2Q^{19/2} + 2Q^{12} + O(Q^{29/2})\right),$

(6.4.8)
$$\left(\frac{2}{5 - \sqrt{5}} A_1 A_{34} + \frac{2}{5 + \sqrt{5}} B_1 B_{34}\right)^4 = 1 + 4Q^{5/2} + 10Q^5 + 20Q^{15/2} + 39Q^{10} + O(Q^{25/2}),$$

and
(6.4.9)
$$\left(\frac{2}{5 - \sqrt{5}} A_1 A_{34} + \frac{2}{5 + \sqrt{5}} B_1 B_{34}\right)^2 = 1 + 2Q^{5/2} + 3Q^5 + 4Q^{15/2} + 7Q^{10} + O(Q^{25/2}).$$

Hence, from (6.4.4), (6.4.6), and (6.4.8),
(6.4.10)
$$U^4 V^4 = e^{-2t} Q^{-9}\left(Q^4 + 4Q^{13/2} + 10Q^9 + 16Q^{23/2} + 27Q^{14} + O(Q^{33/2})\right),$$

and from (6.4.5), (6.4.7), and (6.4.9),

(6.4.11) $qU^2 V^2 = e^{-2t} Q^{-9/2}\left(Q^2 + 2Q^{9/2} + 3Q^7 + 2Q^{19/2} + 5Q^{12} + O(Q^{29/2})\right).$

Hence, by (6.4.10) and (6.4.11),
$$U^4 V^4 - qU^2 V^2 = e^{-2t}\left(Q^{-5} + 3Q^{-5/2} + 8 + 13Q^{5/2} + 25Q^5 + O(Q^{15/2})\right),$$

which, since $v = Q^{5/2}$, agrees with the asymptotic expansion in (6.4.1).

6.5. "Proof" of Entry 3.29

Let $v = e^{-2\pi^2/(23t)}$. Then, from (6.1.16) and (6.1.17), respectively, we find that

(6.5.1) $$\chi(-q)\chi(-q^{23}) = 2\exp\left(-\frac{2\pi^2}{23t} - t\right)\left(1 + v + v^2 + 2v^3 + \cdots\right)$$

and

(6.5.2) $$\frac{2q^2}{\chi(-q)\chi(-q^{23})} = \exp\left(\frac{2\pi^2}{23t} - t\right)\left(1 - v - v^3 + v^4 - v^5 + \cdots\right).$$

Hence, by (6.5.1) and (6.5.2),
(6.5.3)
$$\chi(-q)\chi(-q^{23}) + q + \frac{2q^2}{\chi(-q)\chi(-q^{23})} = e^{-t}\left(v^{-1} + 0 + 2v + v^2 + 3v^3 + 3v^4 + \cdots\right).$$

Next, we calculate the asymptotic expansions of the products of Rogers–Ramanujan functions. For the number of correct terms in the first asymptotic expansion, we need the first two terms from the asymptotic expansion of each of $G(q^2)$ and $H(q^2)$. To that end, from (6.1.1), (6.1.2), (6.1.4), and (6.1.5), we deduce that

(6.5.4) $$G(q^2)G(q^{23}) + q^5 H(q^2)H(q^{23}) = \exp\left(\frac{5\pi^2}{6 \cdot 23t} - \frac{5t}{12}\right) \\ \times \left(1 + Q^5 + Q^{10} + Q^{25/2} + Q^{15} + \cdots\right),$$

where $Q = e^{-4\pi^2/(5 \cdot 23t)}$, and

(6.5.5) $$G(q^{46})H(q) - q^9 G(q)H(q^{46}) = \exp\left(\frac{47\pi^2}{15 \cdot 46t} - \frac{7t}{12}\right) \\ \times \left(Q^{1/2} + Q^{11/2} + Q^8 + Q^{21/2} + Q^{13} + \cdots\right).$$

Thus, (6.5.4) and (6.5.5) yield
$$\{G(q^2)G(q^{23}) + q^5 H(q^2)H(q^{23})\}\{G(q^{46})H(q) - q^9 G(q)H(q^{46})\}$$
$$= \exp\left(\frac{12\pi^2}{5 \cdot 23t} - t\right)\left(1 + Q^5 + Q^{10} + Q^{25/2} + Q^{15} + \cdots\right)$$
$$\times \left(Q^{1/2} + Q^{11/2} + Q^8 + Q^{21/2} + Q^{13} + \cdots\right)$$
(6.5.6) $$= \exp\left(\frac{2\pi^2}{23t} - t\right)\left(1 + 2Q^5 + Q^{15/2} + 3Q^{10} + 3Q^{25/2} + \cdots\right).$$

Since $v = Q^{5/2}$, we see that the asymptotic expansions in (6.5.3) and (6.5.6) agree through the terms that have been calculated.

6.6. "Proof" of Entry 3.35

It will be convenient to use the notation of Subsection 6.4. From (6.1.1) and (6.1.2),

(6.6.1) $$G(q)G(q^{94}) + q^{19}H(q)H(q^{94})$$
$$= \exp\left(\frac{19\pi^2}{3 \cdot 94t} - \frac{19t}{12}\right)\left(\frac{2}{5 - \sqrt{5}}A_1 A_{94} + \frac{2}{5 + \sqrt{5}}B_1 B_{94}\right)$$

and
(6.6.2)
$$G(q^{47})H(q^2) - q^9 G(q^2)H(q^{47}) = \frac{1}{\sqrt{5}} \exp\left(\frac{49\pi^2}{30 \cdot 47t} - \frac{5t}{12}\right)(A_{47}B_2 - A_2 B_{47}).$$

Set $Q = e^{-4\pi^2/(5 \cdot 94t)}$. Then, from (6.6.1), (6.6.2), (6.1.4), and (6.1.5), we find that
(6.6.3)
$$\{G(q)G(q^{94}) + q^{19}H(q)H(q^{94})\}\{G(q^{47})H(q^2) - q^9 G(q^2)H(q^{47})\}$$
$$= \frac{1}{\sqrt{5}} \exp\left(\frac{48\pi^2}{5 \cdot 94t} - 2t\right)\left(\frac{2}{5-\sqrt{5}}A_1 A_{94} + \frac{2}{5+\sqrt{5}}B_1 B_{94}\right)(A_{47}B_2 - A_2 B_{47})$$
$$= e^{-2t}Q^{-12}\left(1 + Q^5 + Q^{10} + Q^{15} + 2Q^{20} + 2Q^{25} + 3Q^{30} + 3Q^{35} + 4Q^{40} + \cdots\right)$$
$$\times \left(Q^2 + Q^{22} + Q^{32} + Q^{42} + Q^{52} + \cdots\right)$$
$$= e^{-2t}\left(Q^{-10} + Q^{-5} + 1 + Q^5 + 3Q^{10} + 3Q^{15} + 5Q^{20} + 5Q^{25} + 8Q^{30} + \cdots\right).$$

Next, by (6.1.16), with $v = e^{-2\pi^2/(47t)}$,
$$\chi(-q)\chi(-q^{47}) = 2\exp\left(-\frac{4\pi^2}{47t} - 2t\right)$$
(6.6.4)
$$\times (1 + v + v^2 + 2v^3 + 2v^4 + 3v^5 + 4v^6 + \cdots).$$

We next invoke (6.1.17) to deduce that
$$\frac{2q^4}{\chi(-q)\chi(-q^{47})} = \exp\left(\frac{4\pi^2}{47t} - 2t\right)$$
(6.6.5)
$$\times (1 - v - v^3 + v^4 - v^5 + v^6 - v^7 + 2v^8 - 2v^9 + 2v^{10} + \cdots).$$

Hence, noting that $v = Q^5$, we find from (6.6.4) and (6.6.5) that

(6.6.6) $\chi(-q)\chi(-q^{47}) + 2q^2 + \dfrac{2q^4}{\chi(-q)\chi(-q^{47})}$
$$= e^{-2t}\left(Q^{-10} - Q^{-5} + 2 - Q^5 + 3Q^{10} + Q^{15} + 3Q^{20} + 3Q^{25} + 6Q^{30} + \cdots\right)$$

Furthermore, from (6.6.4), (6.6.5), and *Mathematica*,
(6.6.7)
$$q\sqrt{4\chi(-q)\chi(-q^{47}) + 9q^2 + \frac{8q^4}{\chi(-q)\chi(-q^{47})}} = e^{-2t}Q^{-5}$$
$$\times \sqrt{4 - 4v + 9v^2 - 4v^3 + 12v^4 + 4v^5 + 12v^6 + 12v^7 + 24v^8 + 16v^9 + 40v^{10} + \cdots}$$
$$= e^{-2t}\left(2Q^{-5} - 1 + 2Q^5 + 2Q^{15} + 2Q^{20} + 2Q^{25} + 2Q^{30} + \cdots\right).$$

Adding (6.6.7) and (6.6.8), we conclude that
(6.6.8)
$$\chi(-q)\chi(-q^{47}) + 2q^2 + \frac{2q^4}{\chi(-q)\chi(-q^{47})} + q\sqrt{4\chi(-q)\chi(-q^{47}) + 9q^2 + \frac{8q^4}{\chi(-q)\chi(-q^{47})}}$$
$$= e^{-2t}\left(Q^{-10} + Q^{-5} + 1 + Q^5 + 3Q^{10} + 3Q^{15} + 5Q^{20} + 5Q^{25} + 8Q^{30} + \cdots\right).$$

Noting that (6.6.3) and (6.6.8) agree to the number of calculated terms, we conclude our asymptotic "proof" of Entry 3.35.

CHAPTER 7

New Identities for $G(q)$ and $H(q)$ and Final Remarks

For proofs of the following three theorems giving new identities for $G(q)$ and $H(q)$, see [9].

THEOREM 7.1. *We have*

$$\frac{G(-q^2)G(q^{38}) - q^8 H(-q^2)H(q^{38})}{G(q^{152})H(q^2) - q^{30}G(q^2)H(q^{152})} = \frac{G(q^{38})H(q^8) - q^6 G(q^8)H(q^{38})}{G(q^2)G(-q^{38}) - q^8 H(q^2)H(-q^{38})}$$
(7.1)
$$= \frac{G(q^{19})H(q^4) - q^3 G(q^4)H(q^{19})}{G(q^{76})H(-q) + q^{15}G(-q)H(q^{76})} = \frac{\chi(-q^2)}{\chi(-q^{38})}$$

and

$$\{G(q)G(-q^{19}) - q^4 H(q)H(-q^{19})\}\{G(-q)G(q^{19}) - q^4 H(-q)H(q^{19})\}$$
$$= \{G(q^{19})H(q^4) - q^3 H(q^{19})G(q^4)\}\{G(q^{76})H(q) - q^{15}H(q^{76})G(q)\}$$
(7.2) $\quad = G(q^2)G(q^{38}) + q^8 H(q^2)H(q^{38}).$

THEOREM 7.2. *Let*

(7.3) $\qquad B(q) := G(q^{12})H(-q^7) + qG(-q^7)H(q^{12}),$

(7.4) $\qquad C(q) := G(q)G(q^{84}) + q^{17}H(q)H(q^{84}),$

(7.5) $\qquad V(q) := H(-q)G(q^{21}) + q^4 G(-q)H(q^{21}),$

(7.6) $\qquad W(q) := G(q^4)G(q^{21}) + q^5 H(q^4)H(q^{21}),$

(7.7) $\qquad Z(q) := H(q^3)G(q^{28}) - q^5 G(q^3)H(q^{28}),$

(7.8) $\qquad Y(q) := G(q^3)G(-q^7) - q^2 H(q^3)H(-q^7).$

Then,

(7.9) $\qquad \dfrac{C(q^2)}{Y(-q^2)} = \dfrac{V(-q^2)}{B(-q^2)} = \dfrac{C(q)}{B(q)} = \dfrac{f(-q^{12})f(-q^{14})}{f(-q^2)f(-q^{84})}$

and

(7.10) $\qquad \dfrac{Z(-q)}{W(q)} = \dfrac{Z(q)}{W(-q)} = \dfrac{Y(q^2)}{W(q^2)} = \dfrac{Z(q^2)}{V(q^2)} = \dfrac{f(-q^4)f(-q^{42})}{f(-q^6)f(-q^{28})}.$

S.-S. Huang [19] derived an identity that belongs to the same class of identities as those in Theorem 7.2, but is different from any identity in (7.9) or (7.10). Huang expressed

$$\{G(q^4)G(q^{21}) + q^5 H(q^4)H(q^{21})\}\{G(q)G(q^{84}) + q^{17}H(q)H(q^{84})\}$$

in terms of two quotients, each with 14 functions of the form $f(-q^n)$.

THEOREM 7.3.
$$\frac{G(q)G(-q^{14}) - q^3 H(q)H(-q^{14})}{G(q^7)H(-q^2) + qH(q^7)G(-q^2)} = \frac{G(q^{56})H(q) - q^{11}H(q^{56})G(q)}{G(q^7)G(q^8) + q^3 H(q^7)H(q^8)}$$
(7.11)
$$= \frac{\chi(-q^{14})}{\chi(-q^2)} = \frac{G(q)G(q^{14}) + q^3 H(q)H(q^{14})}{G(-q^7)H(q^2) + qH(-q^7)G(q^2)}.$$

Besides the isolated identity found by Huang [19], we know of only two further sources of new identities for $G(q)$ and $H(q)$ in the literature that are in the spirit of Ramanujan's identities. The first are found in an unpublished doctoral dissertation by S. Robins [30]. His 13 new identities are associated with modular equations of degree not exceeding 7 and were proved using the theory of modular forms. The second group were found by M. Koike [22]. He discovered them using Thompson series and a computer, but he did not prove them.

One might ask if comparable identities hold for functions similar to the Rogers–Ramanujan functions. Indeed, Huang [19] has derived several identities of this type for the Göllnitz–Gordon functions, and Baruah and Bora [3] have found further identities for these functions. H. Hahn [17] has derived a large number of identities for septic analogues of the Rogers–Ramanujan functions.

Bibliography

[1] G. E. Andrews, *Partitions: Yesterday and Today*, The New Zealand Mathematical Society, Wellington, 1979.

[2] G. E. Andrews and D. Hickerson, *Ramanujan's "Lost" Notebook VII: The sixth order mock theta functions*, Adv. Math. **89** (1991), 60–105.

[3] N. D. Baruah and J. Bora *Nonic analogues of the Rogers–Ramanujan functions*, submitted for publication.

[4] N. D. Baruah, J. Bora, and N. Saikia, *Some new proofs of modular identities for Göllnitz–Gordon functions*, Ramanujan J., to appear.

[5] B. C. Berndt, *Ramanujan's Notebooks, Part III*, Springer–Verlag, New York, 1991.

[6] B. C. Berndt, H. H. Chan, S. H. Chan, and W.-C. Liaw, *Cranks and dissections in Ramanujan's lost notebook*, J. Comb. Thy., Ser. A **109** (2005), 91–120.

[7] B. C. Berndt and K. Ono, *Ramanujan's unpublished manuscript on the partition and tau functions with proofs and commentary*, Sém. Lotharingien de Combinatoire **42** (1999), 63 pp.; in *The Andrews Festschrift*, D. Foata and G.-N. Han, eds., Springer–Verlag, Berlin, 2001, pp. 39–110.

[8] B. C. Berndt and R. A. Rankin, *Ramanujan: Letters and Commentary*, American Mathematical Society, Providence, 1995; London Mathematical Society, London, 1995.

[9] B. C. Berndt and H. Yesilyurt, *Ramanujan's forty identities for the Rogers–Ramanujan functions and linear relations between them*, Acta Arith., to appear.

[10] A. J. F. Biagioli, *A proof of some identities of Ramanujan using modular forms*, Glasgow Math. J. **31** (1989), 271–295.

[11] B. J. Birch, *A look back at Ramanujan's notebooks*, Math. Proc. Cambridge Philos. Soc. **78** (1975), 73–79.

[12] R. Blecksmith, J. Brillhart, and I. Gerst, *Some infinite product identities*, Math. Comp. **51** (1988), 301–314.

[13] R. Blecksmith, J. Brillhart, and I. Gerst, *A fundamental modular identity and some applications*, Math. Comp. **61** (1993), 83–95.

[14] D. Bressoud, *Proof and Generalization of Certain Identities Conjectured by Ramanujan*, Ph. D. Thesis, Temple University, 1977.

[15] D. Bressoud, *Some identities involving Rogers–Ramanujan-type functions*, J. London Math. Soc. (2) **16** (1977), 9–18.

[16] H. B. C. Darling, *Proofs of certain identities and congruences enunciated by S. Ramanujan*, Proc. London Math. Soc. (2) **19** (1921), 350–372.

[17] H. Hahn, *Septic analogues of the Rogers–Ramanujan functions*, Acta Arith. **110** (2003), 381–399.

[18] D. Hickerson, *A proof of the mock theta conjectures*, Invent. Math. **94** (1988), 639–660.

[19] S.-S. Huang, *On modular relations for the Göllnitz–Gordon functions with applications to partitions*, J. Number Thy. **68** (1998), 178–216.

[20] S.-Y. Kang, *Some theorems on the Rogers–Ramanujan continued fraction and associated theta function identities in Ramanujan's lost notebook*, Ramanujan J. **3** (1999), 91–111.

[21] M. Katsurada, *Asymptotic expansions of certain q-series and a formula of Ramanujan for specific values of the Riemann zeta function*, Acta Arith. **107** (2003), 269–298.

[22] M. Koike, *Thompson series and Ramanujan's identities*, in *Galois Theory and Modular Forms*, K. Hashimoto, K. Miyake, and H. Nakamura, eds., Kluwer, Dordrecht, 2003, pp. 367–374.

[23] J. Lehner, *A partition function connected with the modulus five*, Duke Math. J. **8** (1941), 631–655.

[24] L. J. Mordell, *Note on certain modular relations considered by Messrs. Ramanujan, Darling and Rogers*, Proc. London Math. Soc. (2) **20** (1922), 408–416.

[25] S. Ramanujan, *Proof of certain identities in combinatory analysis*, Proc. Cambridge Philos. Soc. **19** (1919), 214–216.

[26] S. Ramanujan, *Algebraic relations between certain infinite products*, Proc. London Math. Soc. **2** (1920), p. xviii.

[27] S. Ramanujan, *Collected Papers*, Cambridge University Press, Cambridge, 1927; reprinted by Chelsea, New York, 1962; reprinted by the American Mathematical Society, Providence, RI, 2000.

[28] S. Ramanujan, *Notebooks* (2 volumes), Tata Institute of Fundamental Research, Bombay, 1957.

[29] S. Ramanujan, *The Lost Notebook and Other Unpublished Papers*, Narosa, New Delhi, 1988.

[30] S. Robins, *Arithmetic Properties of Modular Forms*, Ph. D. Thesis, University of California at Los Angeles, 1991.

[31] L. J. Rogers, *Second memoir on the expansion of certain infinite products*, Proc. London Math. Soc. **25** (1894), 318–343.

[32] L. J. Rogers, *On a type of modular relation*, Proc. London Math. Soc. **19** (1921), 387–397.

[33] G. N. Watson, *Theorems stated by Ramanujan* (VII): *Theorems on continued fractions*, J. London Math. Soc. **4** (1929), 39–48.

[34] G. N. Watson, *Proof of certain identities in combinatory analysis*, J. Indian Math. Soc. **20** (1933), 57–69.

Editorial Information

To be published in the *Memoirs*, a paper must be correct, new, nontrivial, and significant. Further, it must be well written and of interest to a substantial number of mathematicians. Piecemeal results, such as an inconclusive step toward an unproved major theorem or a minor variation on a known result, are in general not acceptable for publication.

Papers appearing in *Memoirs* are generally at least 80 and not more than 200 published pages in length. Papers less than 80 or more than 200 published pages require the approval of the Managing Editor of the Transactions/Memoirs Editorial Board.

As of February 28, 2007, the backlog for this journal was approximately 15 volumes. This estimate is the result of dividing the number of manuscripts for this journal in the Providence office that have not yet gone to the printer on the above date by the average number of monographs per volume over the previous twelve months, reduced by the number of volumes published in four months (the time necessary for preparing a volume for the printer). (There are 6 volumes per year, each usually containing at least 4 numbers.)

A Consent to Publish and Copyright Agreement is required before a paper will be published in the *Memoirs*. After a paper is accepted for publication, the Providence office will send a Consent to Publish and Copyright Agreement to all authors of the paper. By submitting a paper to the *Memoirs*, authors certify that the results have not been submitted to nor are they under consideration for publication by another journal, conference proceedings, or similar publication.

Information for Authors

Memoirs are printed from camera copy fully prepared by the author. This means that the finished book will look exactly like the copy submitted.

Initial submission. The AMS uses Centralized Manuscript Processing for initial submissions. Authors should submit a PDF file using the Initial Manuscript Submission form found at www.ams.org/cgi-bin/peertrack/submission.pl, or send one copy of the manuscript to the following address: Centralized Manuscript Processing, MEMOIRS OF THE AMS, 201 Charles Street, Providence, RI 02904-2294 USA. If a paper copy is being forwarded to the AMS, indicate that it is for it Memoirs and include the name of the corresponding author, contact information such as email address or mailing address, and the name of an appropriate Editor to review the paper (see the list of Editors below).

The paper must contain a *descriptive title* and an *abstract* that summarizes the article in language suitable for workers in the general field (algebra, analysis, etc.). The *descriptive title* should be short, but informative; useless or vague phrases such as "some remarks about" or "concerning" should be avoided. The *abstract* should be at least one complete sentence, and at most 300 words. Included with the footnotes to the paper should be the 2000 *Mathematics Subject Classification* representing the primary and secondary subjects of the article. The classifications are accessible from www.ams.org/msc/. The list of classifications is also available in print starting with the 1999 annual index of *Mathematical Reviews*. The Mathematics Subject Classification footnote may be followed by a list of *key words and phrases* describing the subject matter of the article and taken from it. Journal abbreviations used in bibliographies are listed in the latest *Mathematical Reviews* annual index. The series abbreviations are also accessible from www.ams.org/publications/. To help in preparing and verifying references, the AMS offers MR Lookup, a Reference Tool for Linking, at www.ams.org/mrlookup/.

Electronically prepared manuscripts. The AMS encourages electronically prepared manuscripts, with a strong preference for $\mathcal{A}_{\mathcal{M}}\mathcal{S}$-LaTeX. To this end, the Society has prepared $\mathcal{A}_{\mathcal{M}}\mathcal{S}$-LaTeX author packages for each AMS publication. Author packages include instructions for preparing electronic manuscripts, samples, and a style file that generates the particular design specifications of that publication series. Though $\mathcal{AMS}$-LaTeX is the highly preferred format of TeX, author packages are also available in $\mathcal{AMS}$-TeX.

Authors may retrieve an author package from the AMS website starting from `www.ams.org/tex/` or via FTP to `ftp.ams.org` (login as `anonymous`, enter username as password, and type `cd pub/author-info`). The *AMS Author Handbook* and the *Instruction Manual* are available in PDF format following the author packages link from `www.ams.org/tex/`. The author package can also be obtained free of charge by sending email to `tech-support@ams.org` (Internet) or from the Publication Division, American Mathematical Society, 201 Charles St., Providence, RI 02904-2294, USA. When requesting an author package, please specify $\mathcal{AMS}$-LaTeX or $\mathcal{AMS}$-TeX and the publication in which your paper will appear. Please be sure to include your complete mailing address.

After acceptance. The final version of the electronic file should be sent to the Providence office (this includes any TeX source file, any graphics files, and the DVI or PostScript file) immediately after the paper has been accepted for publication.

Before sending the source file, be sure you have proofread your paper carefully. The files you send must be the EXACT files used to generate the proof copy that was accepted for publication. For all publications, authors are required to send a printed copy of their paper, which exactly matches the copy approved for publication, along with any graphics that will appear in the paper.

Accepted electronically prepared files can be submitted via the web at `www.ams.org/submit-book-journal/`, sent via FTP, or sent on CD-Rom or diskette to the Electronic Prepress Department, American Mathematical Society, 201 Charles Street, Providence, RI 02904-2294 USA. TeX source files, DVI files, and PostScript files can be transferred over the Internet by FTP to the Internet node `ftp.ams.org` (130.44.1.100). When sending a manuscript electronically via CD-Rom or diskette, please be sure to include a message identifying the paper as a Memoir.

Electronically prepared manuscripts can also be sent via email to `pub-submit@ams.org` (Internet). In order to send files via email, they must be encoded properly. (DVI files are binary and PostScript files tend to be very large.)

Electronic graphics. Comprehensive instructions on preparing graphics are available at `www.ams.org/jourhtml/`. A few of the major requirements are given here.

Submit files for graphics as EPS (Encapsulated PostScript) files. This includes graphics originated via a graphics application as well as scanned photographs or other computer-generated images. If this is not possible, TIFF files are acceptable as long as they can be opened in Adobe Photoshop or Illustrator. No matter what method was used to produce the graphic, it is necessary to provide a paper copy to the AMS.

Authors using graphics packages for the creation of electronic art should also avoid the use of any lines thinner than 0.5 points in width. Many graphics packages allow the user to specify a "hairline" for a very thin line. Hairlines often look acceptable when proofed on a typical laser printer. However, when produced on a high-resolution laser imagesetter, hairlines become nearly invisible and will be lost entirely in the final printing process.

Screens should be set to values between 15% and 85%. Screens which fall outside of this range are too light or too dark to print correctly. Variations of screens within a graphic should be no less than 10%.

Inquiries. Any inquiries concerning a paper that has been accepted for publication should be sent to `memo-query@ams.org` or directly to the Electronic Prepress Department, American Mathematical Society, 201 Charles St., Providence, RI 02904-2294 USA.

Editors

This journal is designed particularly for long research papers, normally at least 80 pages in length, and groups of cognate papers in pure and applied mathematics. Papers intended for publication in the *Memoirs* should be addressed to one of the following editors. The AMS uses Centralized Manuscript Processing for initial submissions to AMS journals. Authors should follow instructions listed on the Initial Submission page found at www.ams.org/memo/memosubmit.html.

Algebra to ALEXANDER KLESHCHEV, Department of Mathematics, University of Oregon, Eugene, OR 97403-1222; email: ams@noether.uoregon.edu

Algebra and its application to MINA TEICHER, Emmy Noether Research Institute for Mathematics, Bar-Ilan University, Ramat-Gan 52900, Israel; email: teicher@macs.biu.ac.il

Algebraic geometry to DAN ABRAMOVICH, Department of Mathematics, Brown University, Box 1917, Providence, RI 02912; email: amsedit@math.brown.edu

Algebraic number theory to V. KUMAR MURTY, Department of Mathematics, University of Toronto, 100 St. George Street, Toronto, ON M5S 1A1, Canada; email: murty@math.toronto.edu

Algebraic topology to ALEJANDRO ADEM, Department of Mathematics, University of British Columbia, Room 121, 1984 Mathematics Road, Vancouver, British Columbia, Canada V6T 1Z2; email: adem@math.ubc.ca

Combinatorics to JOHN R. STEMBRIDGE, Department of Mathematics, University of Michigan, Ann Arbor, Michigan 48109-1109; email: FRS@umich.edu

Complex analysis and harmonic analysis to ALEXANDER NAGEL, Department of Mathematics, University of Wisconsin, 480 Lincoln Drive, Madison, WI 53706-1313; email: nagel@math.wisc.edu

Differential geometry and global analysis to LISA C. JEFFREY, Department of Mathematics, University of Toronto, 100 St. George St., Toronto, ON Canada M5S 3G3; email: jeffrey@math.toronto.edu

Dynamical systems and ergodic theory to AMIE WILKINSON, Department of Mathematics, Northwestern University, 2033 Sheridan Road, Evanston, IL 60208-2730; email: transactions@math.northwestern.edu

Functional analysis and operator algebras to DIMITRI SHLYAKHTENKO, Department of Mathematics, University of California, Los Angeles, CA 90095; email: shlyakht@math.ucla.edu

Geometric analysis to WILLIAM P. MINICOZZI II, Department of Mathematics, Johns Hopkins University, 3400 N. Charles St., Baltimore, MD 21218; email: trans@math.jhu.edu

Geometric analysis to MLADEN BESTVINA, Department of Mathematics, University of Utah, 155 South 1400 East, JWB 233, Salt Lake City, Utah 84112-0090; email: bestvina@math.utah.edu

Harmonic analysis, representation theory, and Lie theory to ROBERT J. STANTON, Department of Mathematics, The Ohio State University, 231 West 18th Avenue, Columbus, OH 43210-1174; email: stanton@math.ohio-state.edu

Logic to STEFFEN LEMPP, Department of Mathematics, University of Wisconsin, 480 Lincoln Drive, Madison, Wisconsin 53706-1388; email: lempp@math.wisc.edu

Partial differential equations to GUSTAVO PONCE, Department of Mathematics, South Hall, Room 6607, University of California, Santa Barbara, CA 93106; email: ponce@math.ucsb.edu

Partial differential equations and dynamical systems to PETER POLACIK, School of Mathematics, University of Minnesota, Minneapolis, MN 55455; email: polacik@math.umn.edu

Probability and statistics to KRZYSZTOF BURDZY, Department of Mathematics, University of Washington, Box 354350, Seattle, Washington 98195-4350; email: burdzy@math.washington.edu

Real analysis and partial differential equations to DANIEL TATARU, Department of Mathematics, University of California, Berkeley, Berkeley, CA 94720; email: tataru@math.berkeley.edu

All other communications to the editors should be addressed to the Managing Editor, ROBERT GURALNICK, Department of Mathematics, University of Southern California, Los Angeles, CA 90089-1113; email: guralnic@math.usc.edu.

Titles in This Series

883 **Apostolos Beligiannis and Idun Reiten,** Homological and homotopical aspects of torsion theories, 2007

882 **Lars Inge Hedberg and Yuri Netrusov,** An axiomatic approach to function spaces, spectral synthesis, and Luzin approximation, 2007

881 **Tao Mei,** Operator valued Hardy spaces, 2007

880 **Bruce C. Berndt, Geumlan Choi, Youn-Seo Choi, Heekyoung Hahn, Boon Pin Yeap, Ae Ja Yee, Hamza Yesilyurt, and Jinhee Yi,** Ramanujan's forty identities for the Rogers-Ramanujan functions, 2007

879 **O. García-Prada, P. B. Gothen, and V. Muñoz,** Betti numbers of the moduli space of rank 3 parabolic Higgs bundles, 2007

878 **Alessandra Celletti and Luigi Chierchia,** KAM stability and celestial mechanics, 2007

877 **María J. Carro, José A. Raposo, and Javier Soria,** Recent developments in the theory of Lorentz spaces and weighted inequalities, 2007

876 **Gabriel Debs and Jean Saint Raymond,** Borel liftings of Borel sets: Some decidable and undecidable statements, 2007

875 **C. Krattenthaler and T. Rivoal,** Hypergéométrie et fonction zêta de Riemann, 2007

874 **Sonia Natale,** Semisolvability of semisimple Hopf algebras of low dimension, 2007

873 **A. J. Duncan,** Exponential genus problems in one-relator products of groups, 2007

872 **Anthony V. Geramita, Tadahito Harima, Juan C. Migliore, and Yong Su Shin,** The Hilbert function of a level algebra, 2007

871 **Pascal Auscher,** On necessary and sufficient conditions for L^p-estimates of Riesz transforms associated to elliptic operators on $\mathbb{R}^n$ and related estimates, 2007

870 **Takuro Mochizuki,** Asymptotic behaviour of tame harmonic bundles and an application to pure twistor D-modules, Part 2, 2007

869 **Takuro Mochizuki,** Asymptotic behaviour of tame harmonic bundles and an application to pure twistor D-modules, Part 1, 2007

868 **Gelu Popescu,** Entropy and multivariable interpolation, 2006

867 **Vilmos Totik,** Metric properties of harmonic measures, 2006

866 **William Craig,** Semigroups underlying first-order logic, 2006

865 **Nathanial P. Brown,** Invariant means and finite representation theory of $C*$-algebras, 2006

864 **John M. Lee,** Fredholm operators and Einstein metrics on conformally compact manifolds, 2006

863 **M. Lübke and A. Teleman,** The Universal Kobayashi-Hitchin correspondence on Hermitian manifolds, 2006

862 **Alberto Canonaco,** The Beilinson complex and canonical rings of irregular surfaces, 2006

861 **Leon A. Takhtajan and Lee-Peng Teo,** Weil-Petersson metric on the universal Teichmüller space, 2006

860 **Thomas M. Fiore,** Pseudo limits, biadjoints and pseudo algebras: Categorical foundations of conformal field theory, 2006

859 **N. Arcozzi, R. Rochberg, and E. Sawyer,** Carleson measures and interpolating sequences for Besov spaces on complex balls, 2006

858 **Enrico Valdinoci, Berardino Sciunzi, and Vasile Ovidiu Savin,** Flat level set regularity of p-Laplace phase transitions, 2006

857 **Donatella Danielli, Nocola Garofalo, and Duy-Minh Nhieu,** Non-doubling Ahlfors measures, perimeter measures, and the characterization of the trace spaces of Sobolev functions in Carnot-Carathéodory spaces, 2006

856 **Vladimir Bolotnikov and Harry Dym,** On boundary interpolation for matrix valued Schur functions, 2006

TITLES IN THIS SERIES

855 **Yevgenia Kashina, Yorck Sommerhäuser, and Yongchang Zhu,** On higher Frobenius-Schur indicators, 2006

854 **Noam Greenberg,** The role of true finiteness in the admissible recursively enumerable degrees, 2006

853 **Joachim Krieger,** Stability of spherically symmetric wave maps, 2006

852 **Viorel Barbu, Irena Lasiecka, and Roberto Triggiani,** Tangential boundary stabilization of Navier-Stokes equations, 2006

851 **Jie Wu,** On maps from loop suspensions to loop spaces and the shuffle relations on the Cohen groups, 2006

850 **Siegfried Echterhoff, S. Kaliszewski, John Quigg, and Iain Raeburn,** A categorical approach to imprimitivity theorems for C^*-dynamical systems, 2006

849 **Katsuhiko Kuribayashi, Mamoru Mimura, and Tetsu Nishimoto,** Twisted tensor products related to the cohomology of the classifying spaces of loop groups, 2006

848 **Bob Oliver,** Equivalences of classifying spaces completed at the prime two, 2006

847 **Eric T. Sawyer and Richard L. Wheeden,** Hölder continuity of weak solutions to subelliptic equations with rough coefficients, 2006

846 **Victor Beresnevich, Detta Dickinson, and Sanju Velani,** Measure theoretic laws for lim–sup sets, 2006

845 **Ehud Friedgut, Vojtech Rödl, Andrzej Ruciński, and Prasad V. Tetali,** A Sharp threshold for random graphs with a monochromatic triangle in every edge coloring, 2006

844 **Amadeu Delshams, Rafael de la Llave, and Tere M. Seara,** A geometric mechanism for diffusion in Hamiltonian systems overcoming the large gap problem: Heuristics and rigorous verification on a model, 2006

843 **Denis V. Osin,** Relatively hyperbolic groups: Intrinsic geometry, algebraic properties, and algorithmic problems, 2006

842 **David P. Blecher and Vrej Zarikian,** The calculus of one-sided M-ideals and multipliers in operator spaces, 2006

841 **Enrique Artal Bartolo, Pierrette Cassou-Noguès, Ignacio Luengo, and Alejandro Melle Hernández,** Quasi-ordinary power series and their zeta functions, 2005

840 **Sławomir Kołodziej,** The complex Monge-Ampère equation and pluripotential theory, 2005

839 **Mihai Ciucu,** A random tiling model for two dimensional electrostatics, 2005

838 **V. Jurdjevic,** Integrable Hamiltonian systems on complex Lie groups, 2005

837 **Joseph A. Ball and Victor Vinnikov,** Lax-Phillips scattering and conservative linear systems: A Cuntz-algebra multidimensional setting, 2005

836 **H. G. Dales and A. T.-M. Lau,** The second duals of Beurling algebras, 2005

835 **Kiyoshi Igusa,** Higher complex torsion and the framing principle, 2005

834 **Ken'ichi Ohshika,** Kleinian groups which are limits of geometrically finite groups, 2005

833 **Greg Hjorth and Alexander S. Kechris,** Rigidity theorems for actions of product groups and countable Borel equivalence relations, 2005

832 **Lee Klingler and Lawrence S. Levy,** Representation type of commutative Noetherian rings III: Global wildness and tameness, 2005

831 **K. R. Goodearl and F. Wehrung,** The complete dimension theory of partially ordered systems with equivalence and orthogonality, 2005

For a complete list of titles in this series, visit the
AMS Bookstore at **www.ams.org/bookstore/**.